WIRED for GREED

WIRED for GREED

✦

The Shocking Truth about America's Electric Utilities

Joe Seeber
with Jim Moore

iUniverse, Inc.
New York Lincoln Shanghai

WIRED for GREED
The Shocking Truth about America's Electric Utilities

Copyright © 2005 by Joe Seeber

iUniverse books may be ordered through booksellers or by contacting:

iUniverse
2021 Pine Lake Road, Suite 100
Lincoln, NE 68512
www.iuniverse.com
1-800-Authors (1-800-288-4677)

ISBN-13: 978-0-595-35744-4 (pbk)
ISBN-13: 978-0-595-80224-1 (ebk)
ISBN-10: 0-595-35744-X (pbk)
ISBN-10: 0-595-80224-9 (ebk)

Printed in the United States of America

Contents

PREFACE

Jim Moore and I have been working on this manuscript off and on since 1998. During that time, many events in both my business and the electric utility industry have helped shape its content.

The notion of writing this book first crossed my mind about twenty years ago, and it remained just that—an idea—for a decade afterward. The original idea occurred to me some years after I had met a utility executive who headed up the rates department of a major power company. When we first met, this gentleman and I were pulling at opposite ends of the rope in a dispute with his power company. He had spent his entire career working for the utility; he also held influential positions in the International Electric and Electronics Engineers (IEEE) organization, which is widely regarded as the leading authority on the electric power industry.

When my executive friend retired, we came to know each other in a different context. As we became friends, we enjoyed hours of discussions focusing on my perception that the power companies were not doing things "the right way." As we argued our different views, the real kicker was his comment (repeated at the end of many of our discussions) that was something akin to this: "Joe, you have a good point. I just never saw it from the customer's point of view."

His altered mind-set had created within him an entirely new attitude, and I found that in most instances he would eventually agree with me. The problem he faced was one of culture: The environment in which he worked had developed in such a way that unacceptable behavior had become acceptable.

This transformation of the unacceptable into the acceptable is a cultural phenomenon. Perhaps the most striking example in modern times is the mass suicide at the People's Temple in Guyana. Jim Jones and his band of followers exiled themselves from the rest of the world, and their dysfunctional culture evolved to the point where the unthinkable became an accepted response.

Similarly, millions of Jews perished in the Holocaust, a highly mechanized process of slaughter that involved the efforts of thousands. From manufacturing the Zyklon B gas to overseeing the starved inmates who piled the corpses into ovens, the unacceptable became acceptable to some in the warped culture of Nazi Germany.

These are extreme examples. Certainly I have no wish to equate electric utilities with the Nazis. My point is that cultures do evolve to the point where unacceptable behavior becomes the norm and is accepted as standard practice. The individuals who perpetuate these misdeeds, such as my friend the former utility executive, are often unaware of the unacceptable aspects of corporate behavior.

We, similarly, are unaware because our perspective of the electric utilities has been molded by the power companies themselves, according to their own view. My friend and his colleagues, past and present, are the unwitting victims of that mind-set. It is almost impossible to find an independent perspective on the electric power industry. Even regulators and the legislative bodies to which they are responsible are not without substantial bias, and most have been co-opted for decades by the utilities themselves.

The resulting sculpting of public opinion about the electric power industry—and the opinion of industry insiders—is the result of decades of careful conditioning. This book offers a less scripted, more accurate view of the industry and how it works. While the electric utility industry itself may be morally bankrupt, people like my old friend are still responsible for its future course. Let us hope that their enlightened views will eventually hold sway, for the betterment of us all.

Joe Seeber
Waco, Texas
Spring 2005

INTRODUCTION

Conventional wisdom says:

If it walks like a duck,
quacks like a duck,
and acts like a duck,
then it is a duck.

Following this logic, we can safely say that a business that has a monopoly will act like a monopoly.

For most of the country, deregulation is a contrived myth. Consequently, electric monopolies continue to act like what they are—and what they always have been.

The giant electric companies still own the transmission and distribution apparatus. Deregulation itself is in a state of flux; the final form of the concept has yet to be determined.

Over the past quarter of a century, our experience has taught us two things:

First, electric utilities do all the things monopolies do—all the things that make monopolies so objectionable.

Second, the general public doesn't know about monopolistic practices, doesn't want to admit to the misbehavior, and is afraid to do anything to stop it.

As defined herein, "the general public" includes regulators and politicians.

This book is our attempt to examine this curious circumstance.

The Light Standard

The weather system roaring through New Orleans on the afternoon of March 18, 1996, was not a hurricane or even a tropical storm, but it was enough, with sustained winds of nearly 60 miles per hour, to drop a 626-pound metal streetlight fixture onto Nathaniel Joseph's head.

Mr. Joseph was in the process of closing his family business, Mitchell Fruit Stand, when the corroded light pole buckled and collapsed. The light standard crashed to the sidewalk below, pulling down the canopy of the business and striking Mr. Joseph. Knocked to the ground by the force of the impact, the business owner was severely injured. He underwent several back surgeries and will require medical care for the rest of his life.

Mr. Joseph and his wife, Kecia, filed suit against Entergy New Orleans, the electric utility responsible for the maintenance of the city's light standards. After a bench trial, the court found Entergy totally liable for Mr. Joseph's injuries and awarded the couple just over $3 million. That singular event should have been an issue only between those who were supposed to maintain the corroded light pole and Mr. Joseph, but, as you will see, the repercussions from this incident rippled far beyond the parties involved.

Cooking Books, Cajun Style

Entergy is the only Fortune 500 company headquartered in downtown New Orleans. Indeed, the Entergy building is something of a downtown landmark, towering over the French Quarter just a few blocks away. The Louisiana Superdome is two blocks to the southwest. Mother's, a unique New Orleans restaurant, is four blocks in the opposite direction.

We know Mother's well, for we relied on the homemade Cajun-style cuisine for our daily sustenance while we spent months working for the City of New Orleans. Our job involved auditing Entergy's financial records related to streetlights, such as the one that fell on Mr. Joseph.

All of the city's light standards—the poles like the one that struck Mr. Joseph—are actually owned by the City of New Orleans. For most of the twentieth century, the City of New Orleans paid Entergy to maintain the streetlights. Entergy was, and, as far as we know, still is, claiming the lights as an $8 million to

$12 million asset on behalf of Entergy New Orleans. Quite simply, the utility has been misrepresenting its financial condition. While $8 million to $12 million may be a drop in the bucket to a giant company like Entergy, the same amount is a substantial consideration for the Entergy New Orleans subsidiary. We believe Entergy is on the shakiest of financial footings, and, as you will see, it is not just our own experience that tells us so.

Entergy New Orleans is one of the smallest subsidiary companies owned by the Entergy Corporation. The New Orleans operation provides power only within the city limits. Our audit work was completed in 1994, a couple of years before Mr. Joseph had his encounter with one of the light poles. Our audit demonstrated that the city had bought and paid for all of the streetlights—in spite of the fact that Entergy was carrying between $8 million and $12 million in streetlights on its corporate balance sheet. Imagine selling an automobile and then going to the bank to get a loan using the auto as collateral. The process requires you to tell the bank you still own the car. That is effectively what Entergy did. In August 2003 the firm went to Wall Street to borrow $100 million, using assets as collateral. Listed among those assets were the New Orleans streetlights, which Entergy did not own. In other words, Entergy misrepresented its financial condition, claiming to own assets that did not legitimately belong to the firm.

Seventeen Million Reasons

The key word here is "legitimately." We cannot hold Entergy to a legitimate standard of behavior because the utility has often demonstrated that it possesses none. Over the years we have recovered millions of dollars in Entergy overcharges for our clients. As early as 1988, we discovered that Entergy had overbilled Texas's Sam Houston State University more than $155,000. The next year we recovered nearly $60,000 for the City of Huntsville, Texas, and a similar amount for Lamar University—both of which were Entergy clients.

In 1990 we recovered nearly $100,000 in overcharges for the City of Beaumont and more than $31,000 for the City of Orange. Both are Texas cities served by Entergy subsidiaries.

Two years later we found that Entergy New Orleans had overcharged the city's water and sewer board more than $1 million. The next year we recovered $400,000 for the city itself, and in 1994 we reclaimed $6 million in overcharges for New Orleans streetlights that were either nonfunctional or nonexistent.

Much of that amount was billed as maintenance and repairs that were never done. The $6 million was the amount for which the city settled; the actual overcharge, we believe, was nearly double that.

We will give you detailed information about some of these episodes later in this book, but it is important to understand at the outset that the heavy hits Entergy sustained on behalf of overbilled clients did not stop there:

- In 1995 we recovered $80,000 for the CNG Towers complex.
- In 1996 we recovered $70,000 for the New Orleans Centre.
- In 1998 we recovered $280,000 for the Texas Department of Transportation.
- In 2000 we recovered more than $90,000 for Louisiana State University.
- In 2001 we recovered another $1.8 million for the City of New Orleans.

We were far from through with Entergy. In 2002 we recovered $50,000 in Entergy overbillings for the United States Coast Guard, and over the next two years we recovered another $7.427 million for the City of New Orleans.

Suffice it to say the folks at Entergy New Orleans had many good reasons to dislike us . . . more than seventeen million of them!

We were convinced, after our initial engagement with the City of New Orleans in 1994, that Entergy would live up to its agreement with the city and effectively restore streetlights that had deteriorated substantially since the utility stopped repairing the lights in 1987. In 2000, when the city's director of utilities began to talk with us about coming back to town for another audit, our response was less than enthusiastic. Our reasoning was simple: Entergy had learned what was wrong, admitted that the problems existed, and promised to fix them. That, we thought, was that.

The Second Battle of New Orleans

The settlement notwithstanding, the city's director of utilities asked us to offer a proposal for a return engagement. We offered to perform a small fee-based audit that would help us discover if a second audit would be worth our while. When we arrived in New Orleans in late 2000, we found a lot of things had gone wrong. Specifically, the work Entergy had pledged to do with regard to streetlights had not been done. In 1996 Entergy had entered into a new agreement with the city, in which the utility agreed to handle streetlight maintenance and the city agreed to pay for time and materials. When we examined the streetlight billings, we found the following:

- Work had been billed and paid for but not performed.
- Work done for others had been billed to the city.
- Time sheets had been altered to show that workers had spent more time working on streetlights than they had actually spent.
- The city had been billed twice for the same work.

Those were just violations of the 1996 agreement. We discovered that the City of New Orleans and Entergy had entered into another, more recent compact as well. Our audit of the billings delivered under the 1999 contract uncovered the following:

- The new agreement between Entergy and the City of New Orleans was invalid. It had never been submitted to the city council for approval and therefore was not legally executed.
- Entergy had billed the city for streetlights that did not exist.
- Again, Entergy had billed for work not performed.
- Again, Entergy had done work for others and billed the city.
- Again, Entergy had falsified time sheets to increase billings.
- Entergy had not provided periodic reports of work done, which were required by the 1999 contract.
- Entergy had never provided required drawings of the city's underground streetlight circuits.

The 1999 contract included a penalty clause, and fines for Entergy's violations in these areas had grown to several million dollars by the time we began our second audit. Unbelievably, the new city administration chose to waive the penalties. We found it hard to understand why elected officials, who had sworn to uphold the charter of the city and the constitution of the state, had chosen to let Entergy off the hook. Indeed, we could not comprehend the reasoning behind the mayor's appointment of an Entergy attorney—a lawyer with no experience in municipal law—as the new city attorney.

Proof of the Plot

There may be a simple explanation for this, but we suspect collusion. Entergy had friends among the city's new officeholders, and we would later discover that the utility appeared to have some law courts in its rather large pocket as well. Later in this book, you will discover that this is nothing new. We have run into case after case in which elected officials are somehow beholden to the power companies and refuse to take actions that might somehow run counter to the utility's best interests.

Along with the penalties that should have been assessed, our review of the billing records uncovered several million dollars in overcharges. Though our study was not yet complete, the press had heard about our findings and pressed an Entergy vice president for comment. According to him, the overcharges were "just a fabrication by a greedy consultant."

We run an honest, reputable business and do a good job for our clients. Indeed, the very nature of our business requires that we maintain a spotless reputation. So it is hard for us to hear a comment like that and not respond in some way.

As we examined the financial statements of Entergy New Orleans, we were shocked to find that the streetlights—which were clearly owned by the city—were listed as a utility asset. That discovery—and the VP's comment about "a greedy consultant"—led us to request a meeting with Entergy CEO Wayne Leonard. "Mr. Leonard is the CEO of a Fortune 500 company," we were told initially. "Everybody wants to see him, and he's too busy to see you." We then met with one of the company's top-ranking attorneys and told him that Entergy's books were cooked and that we could prove it. That proof, however, would be offered only to the man at the top. Ranting, raving, and stomping about the room, the corporate attorney slapped his hand on the conference table in front of us and repeated that Mr. Leonard would not see us. The lawyer demanded that we tell him what we knew. Of course, we refused.

When the attorney finally realized that we were serious and would not disclose our information, he stormed out of the room. "We're outta here," he yelled to no one in particular.

Two hours later, we received a call. We got an appointment with Mr. Leonard that very afternoon.

The CEO's office occupies much of the top floor of the Entergy building. The view from that office is breathtaking, with the Crescent City sprawling below. The town was built up along a curve in the mighty Mississippi River, and now the urban sprawl stretches away as far as the eye can see. While New Orleans itself, with its bright, garish colors, can be a bit overwhelming, the Entergy CEO's conference room was a muted understatement. The lighting was kept low—so low, in fact, that one could hardly see the expressions on the faces of those across the table. We would have given a lot to turn up the lights just a bit.

The meeting lasted about ninety minutes. Mr. Leonard was made aware of the fact that his subordinate's characterization of our firm as "greedy consultants" was uncalled for and inaccurate. Had we been greedy consultants, we allowed, we would have sold Entergy stock short, disclosed what we knew to the press, and covered the short when the stock fell. We could have made a lot of money that way.

In addition to falsely declaring that the streetlights belonged to the utility, Entergy's books revealed other interesting anomalies: Items that should have been written off as expenses, such as raincoats, saws, batteries, coffee cups, coffee stirrers, aspirin, and motor oil, were capitalized as assets. We have pointed to these items as additional evidence that Entergy New Orleans was bent on inflating its own worth.

Since our meeting with Mr. Leonard, not a single Entergy officer has disputed our findings, except to say that our findings are "somehow not material." We have shared our documentation with Entergy's Audit Committee, the Federal Energy Regulatory Commission, and other parties at interest. Still, not one reputable individual has disputed our assessment of Entergy's shady accounting. Even the firms responsible for auditing Entergy's financial condition have had no comment. Price Waterhouse Coopers (PWC) was Entergy's auditor until 2000; PWC is also the auditor of one of the largest of Entergy's shareholders, AXA Financial Group. Since 2000, Entergy itself has been audited by Deloitte and Touche.

Entergy's legal response, in contrast, has been swift and to the point. The utility filed a motion for a protective order to keep us from disclosing its accounting misdeeds. Since then, Entergy has gone back to court three times to obtain additional protective orders that will shield the company from "greedy consultants" who wish to communicate the fact that the utility's books appear to have been cooked. One of those protective orders in particular sought to restrain us from communicating with the Audit Committee of Entergy's Board of Directors.

Remember the Fruit Stand?

Plainly, these protective orders should not be necessary to protect a utility giant against charges of fraud. Obviously, we have shared our information with many parties, including Entergy itself. In every case we have offered to retract what we have said and apologize if we receive a Letter of Opinion from one of the "Big Four" accounting firms stating that our findings are incorrect. To date, there has been no such letter. We find the lack of response unusual, given the serious nature of the charges we have made. Additionally, we believe that the court-ordered gags effectively violate our First Amendment rights.

In 2000 Entergy named our company as a party at interest in the Joseph case. You will recall that Nathaniel Joseph was struck and seriously injured by a light pole during a spring storm in March 1996. Entergy reasoned that since our firm was consulting with the city about streetlights in 1994, we should have noticed the corroded state of the light standard, anticipated that it might fall or do damage two years hence, and brought the matter to the attention of the utility.

Not that such notice would have done much good. Our streetlight audits had uncovered a good many missed maintenance opportunities during the period in which Entergy was supposed to keep the city's streetlights in good repair. We worked for the city rather than the utility, and the expectation that we could, from our adversarial position, convince the utility to replace one light standard when thousands were begging for attention strikes us as somewhat foolish. Well, it did not seem foolish to a judge; Entergy is looking to our firm to repay the money it had to give Mr. Joseph and his wife.

Entergy's legal action against our company is completely unwarranted and without merit. Indeed, the suit is simply a counterpunch response because we dared to tell the truth about Entergy's financial behavior. Naming us as a party at interest in the Joseph case is not exactly the wildest salvo fired in our war with Entergy, although the action itself is pretty difficult to comprehend with any sense of logic. Indeed, some of the legal remedies sought by the utility have provided us with an extraordinary glimpse of the illogic that appears to grip much of the New Orleans judicial process. A case in point was Entergy's request for a local court to issue a protective order forbidding us to communicate with its Board of Directors' Audit Committee.

Entergy quite properly requested a hearing, and the court agreed to schedule the event. Before the hearing could be held, however, the court issued the requested protective order. When we asked the judge how he could issue such an order without a hearing, he responded that we simple Texans "just don't understand how things work down here." Amazingly enough, the magistrate delivered that line without laughing!

During the months while this manuscript was being written, we were cited twice for contempt of court as we attempted to communicate with members of Entergy's Board of Directors' Audit Committee and others about the utility's accounting alchemy. One has to ask: Is justice blind in the Big Easy, or is justice seeing only what the courts want it to see?

In the post-Enron era, when our society has been shaken to its core by corporate scandals and crooked executives, we feel an obligation to tell what we know. By doing so perhaps we can help prevent another financial disaster and avoid rocking the foundations of American business yet again. Clearly, the pro-utility courts in New Orleans do not see the situation with the same clarity. We believe that the lights in the New Orleans courtrooms, like those in Entergy's executive conference room, are kept low on purpose.

More Trouble in the Big Easy

For all our effort, we may be too late. You see, Entergy has been through all this yet again, in a much more public forum.

In April 2001 Florida Power and Light (FPL) called off a widely anticipated merger with Entergy. Energy analysts said the breakup culminated in one of the most shocking displays of management bitterness they had ever witnessed. We imagine most analysts took a little while to recover from the shock when Entergy and FPL announced that their $15.8 billion deal would not happen. The merger would have created the nation's largest electric company.

Fred Schultz, an analyst with Raymond James and Associates in Houston, told an interviewer that "the Entergy people aired more dirty laundry than I've ever heard on a conference call in ten years. I was sitting here in complete astonishment."

Mr. Schultz's amazement was widely shared, particularly by executives at Florida Power and Light. FPL blamed Entergy for the breakup, saying that Entergy had offered "significantly higher earnings projections than Entergy gave its own board and investment bankers." We had thought Entergy cooked books for equal benefit of all concerned. That, it seemed, was not the case; the Entergy Board of Directors may have gotten something akin to the real story.

Entergy CEO Wayne Leonard said that "corporate cultural differences" began to become apparent two months before the breakup was announced. According to Mr. Leonard, FPL Chairman and CEO James Broadhead called Entergy's management style "chaotic" and told Entergy directors back in March that Leonard was not up to the job of CEO. Mr. Leonard maintained that Broadhead wanted to fire both him and Entergy CFO John Wilder, who is currently the CEO of TXU.

If that is true, we can only applaud Mr. Broadhead's intentions.

FPL spokeswoman Mary Lou Kromer told analysts that her utility did not want to bandy words with Mr. Leonard but that FPL began to lose confidence in Entergy managers when the New Orleans group refused to explain discrepancies in financial forecasts. Around our digs that comment caused quite a chuckle, and now you know why!

An Industry Beset

Electricity is a unique consumable. Except for a brief delay lasting only a few nanoseconds, electricity is produced as it is consumed. In other words, manufacturing of electricity takes place at nearly the same time that electricity is used.

Power providers operate via a demand system. Their production and distribution apparatus must be capable of delivering enough electricity to satisfy consumer demand, on demand. On a hot summer afternoon in Texas, for example, every office air-conditioning system is running full blast, and when the workers go home, the residential air-conditioning units are turned on or thermostats are turned down. This, then, is "peak" demand, the point at which the utilities must supply the most power for a given period of time. During off-peak hours, as much as half of the electric company's generating capacity may be idle.

Most Americans still do not understand electric utilities, and many consumers have only a vague grasp of the intricacies of regulation and deregulation. This is a paradox of sorts. Regulation, in particular, seems easy enough to grasp. The real difficulty lies in understanding how power companies have manipulated the regulators.

Regulators at the state and local levels have traditionally limited the utilities to recovery of costs and a specific margin of profit. It did not take long for the power companies to discover that the best way to increase their profit margin was to increase the costs incurred in the production of electricity. Over the past several decades, electric utilities have become remarkably innovative in developing new and improved methods of increasing their costs of production. In a word, regulators have been effectively co-opted by the power companies.

A case in point: Although we have shared detailed information about Entergy's apparent financial misrepresentation with numerous regulators, the regulators have done nothing. Of course, no matter how long or how loudly one blows the proverbial whistle, there is no guarantee that anyone will listen. Candidly, we expected action of some sort—even if all we received was an allegation of error on our part or a curt dismissal. The Securities and Exchange Commission does maintain a file on the matter, but it has not contacted us in more than a year.

On the Cusp of Change

The hard standards by which we have judged electric utilities over the past century are long outdated. Today, the entire electric industry is in a state of flux. In a very real sense, we have written this book to expose utility malfeasance and promote reform, but it is also intended to dispel confusion and to inform.

Chapter 1 presents a brief history of electric utilities and a description of the internal process of industry regulation.

Chapters 2 and 3 examine deregulation—the perils and potential of utilities competing in the open market.

Finally, Chapter 4 points toward a new beginning for electric utilities—a new vision of the future that requires, perhaps unrealistically, that the utilities learn to adapt to changing circumstances in an honest and decent way, just as other industries have. More than anything else, this fundamental paradigm shift would herald a return to a standard of ethical behavior that we have a right to expect.

We have every reason to expect that the call for adherence to ethical standards will be largely ignored. As you will see, ethical behavior is a quality that is in short supply in the electric utility industry. In decades past, power companies could generally do whatever they wanted—and they did. Because the industry and the world are both in a state of flux, those days will eventually be over. Change will happen, as surely as the bright light of day follows the darkest night. Whether or not the utilities can put themselves back on the road toward honest earnings and ethical standards remains to be seen. Until the power companies offer some proof of their future intentions, their past history gives us every reason to doubt their future conduct.

Is the past really just prologue? From our vantage point, we cannot tell. Change is on the cusp; we cannot predict how the power companies will react to it, but one thing is certain: As deregulation becomes the order of the day in state after state, the easy life enjoyed by America's electric utilities is coming to an abrupt halt. Now companies like Entergy must prepare to atone for past sins with some of their own lifeblood . . . or some will most assuredly drown in it.

1

THE PERIL OF POWER

A Period of Consumer Innocence

The past is prologue. What goes around does indeed come around again. From our perspective, this fundamental belief makes the history of electric utilities in the United States a sometimes tedious but necessary topic. As regulatory agencies face the constraints inherent in managing something that cannot be managed, and as the facade of competition crumbles, the monopolistic status that utilities have enjoyed in decades past is bound to manifest itself again. This time, however, the consumer will likely end up paying far more dearly than ever before for the privilege of buying electricity.

Samuel Insull was Thomas Edison's personal secretary and the founder of Commonwealth Edison, the massive Chicago power company. At the beginning of the twentieth century, Mr. Insull collaborated with local officials to develop a series of tacit understandings that created a model for utilities and the sale of electricity to businesses and residential customers alike.

At that time, several power companies were competing for business in Chicago. Insull realized that the needless duplication of power-generating facilities only raised electricity costs. Several suppliers also meant several sets of distribution wires and generating apparatus. A single seller, Insull reasoned, could sell electricity cheaper because increased production inevitably brought about increased economy.

Mr. Edison's protégé convinced Chicago city officials to grant him a monopoly on the sale of electricity. In return, Insull promised to serve all customers and to allow the city itself to set his rates, so long as those rates guaranteed a reasonable return on his investment.

We will detail Insull's rather sordid story and the history of electric utilities more fully in later chapters. For the moment, let us concern ourselves with the paradox created by Insull's desire for "reasonable return." The regulatory arrangement that ensued eventually became the standard for electric companies across the United States. As more cities and states granted utility companies exclusive franchises, the utilities undertook the development of generation facilities, built distribution networks, and sold electricity to business and residential customers under what amounted to exclusive franchise rights. States, concerned about keeping a tight rein on the reasonable return the utilities sought, created their own regulatory commissions to regulate the electric companies.

This system worked well enough until the end of World War I. Then, in the early 1920s, large holding companies emerged. These holding companies typically owned many utility companies, and they quickly overwhelmed the existing regulatory apparatus. Because the holding companies operated utilities in several different states, they were subject to conflicting regulatory controls. State and local regulatory commissions, which had been developed only to oversee small, local utility operations, were unable to keep track of holding-company revenues. Utility money was easily shifted from one company to another within the same organization or to a cooperating organization headquartered in another state.

As if the situation were not already complicated, it became evident that individual states lacked the jurisdiction to control wholesale electricity transactions across state lines. Indeed, the United States Supreme Court ruled that states could not regulate these interstate sales. As a result, holding companies began shifting assets from state to state in a kind of electric shell game.

Abuses mounted. Unregulated electricity sales became commonplace, and some enterprising providers even sold company stock door-to-door. Behind the scenes, a complex funding system was propping up shaky holding companies. By the late 1920s, the nation's utilities resembled nothing so much as the crashing stock market. Retail customers, held captive by their need for electricity and helpless to regulate the commodity, had no protection from utilities that had run amok.

The harbinger of change was none other than Franklin Delano Roosevelt. Almost immediately after taking office in March 1933, FDR began pressing for legislation aimed at curbing utility abuses and sorting out the complex and confusing electric web. Development of legislation took nearly two years, but the result, the Public Utilities Act of 1935, was intended to solve utility problems.

The PUA was a package of two laws—the Public Utility Holding Company Act and the Federal Power Act. While the Public Utility Holding Company Act dealt with corporate abuse of power, the broad-reaching Federal Power Act was aimed at providing some regulation for the wholesale interstate electricity sales that had thus far avoided the states' legal net.

Passage of the Public Utility Holding Company Act set in motion a modern, more streamlined system for regulating the sale of electricity at both the state and the federal levels. Rates for electricity delivered to retail customers and sold at wholesale were based on utility costs for building and operating transmission, distribution, and generation facilities, in addition to the aforementioned reasonable return on investment.

Customers at every level reaped an almost immediate benefit from this sweeping federal legislation: Rates began to drop. Spurred by an economy of large-scale generation and distribution and aided by significant technological advancement, electric rates steadily declined through the mid-1960s.

From the late 1960s through the early 1970s, however, change began to catch up with the utility industry. Demand for electricity had risen steadily, and utilities were building and expanding plants to handle continued growth. In the early 1970s, the energy crisis forced fuel prices up, and conservation efforts caused demand to drop. Utilities were left with excess capacity and reasoned that customers should pay for it. For the first time, electricity costs began to rise.

At the same time, development of new combustion turbines was well under way. These new units burned natural gas far more efficiently and were far less expensive to build. For the first time, small gas plants could compete effectively with the large coal-burning central stations that had been the hallmark of electric power generation for decades. Additionally, new transmission technologies allowed electricity to be shipped across greater distances, and new switching technology and the advent of computerized control systems made massive regional transmission grids possible.

The federal response was the Public Utility Regulatory Policies Act, adopted in 1978. This act freed alternative generation technologies, such as solar, biomass, or wind, from the regulations set forth in the original Public Utility Holding Company Act and required utilities to buy electricity from these alternative generators at rates equal to the costs that would have been incurred by constructing new facilities.

The result was that wholesale electricity was no longer the sole purview of utility monopolies. Indeed, from the mid-1980s through the mid-1990s, more than half of all new power generation in the United States came from non-utility generators. The Energy Policy Act of 1992 vested in the Federal Electric Regulatory Commission the authority to require utilities to allow their competitors to use utility transmission lines to sell electricity.

Old Samuel Insull probably rolled over in his grave! The industry he had worked to establish and protect was now not only open to competition but also required to actively aid and abet the competing companies. His successors, however, had no intention of going quietly into that competitive night. Over the past several decades, utilities had found dozens of regulatory loopholes and had wielded their monopolistic status quite effectively. The end result saw regulatory cheating become a neo-science, while consumers at every level found themselves put to the grindstone. By the dawn of the new millennium, electricity users great and small were shouldering comparatively massive burdens—all in the name of providing a reasonable rate of return to the empowered and emboldened holding companies nationwide.

The two decades between the end of World War II and the escalation of the Vietnam Conflict were profitable years for America's electric utilities. As the Second World War drew to a close, electric companies nationwide were poised to deliver increasing volumes of their product, and the consumer stood to benefit from progressively lower generating costs. As we will see, this rather euphoric state held sway through the mid-1960s, when a single event forced it to a rather abrupt halt.

As the nation's postwar economy expanded, so did energy consumption. Electricity sales rose at a faster rate than sales of other fuels and energy sources, largely because the price of electricity continued to drop while the prices of other products began to climb. Interestingly, while the big industrial generators of electricity produced more during this period, they contributed less and less to the overall supply. Public utilities made up the difference in supply, and disputes between these publicly held companies and the investor-owned utilities commanded much media attention during the years following 1945.

Why, when the price of other energy sources rose, did the cost of electricity drop? The decline in prices was primarily the result of continued improvements in the way electricity was generated. Large generating stations became permanent fixtures in the urban landscape of the 1950s, and the economy of producing electricity on a large scale became progressively more evident. Through the late 1950s, electric companies were building generating plants with five to ten times the capacity of the plants built during the war years.

With the exception of the year immediately following the end of World War II, the sale of electricity, as measured in kilowatt-hours, grew rapidly each year. Even during the postwar recession, electricity sales surged upward. The electric utility industry could take advantage of declining equipment costs and the increasing economy of generating power on a larger scale. Of course, companies are always willing to expand when they know their new capacity will be utilized, and electric utilities are no different. Confident that the product they generated would be bought, the utilities invested heavily and developed a reserve capacity capable of meeting future demand.

During the 1950s, reserve generating capacity increased, but during the 1960s it declined. During that time, the electric utility industry gobbled up a larger share of the overall market. By 1969, very little electricity was being generated by non-utility sources. The salad years were in evidence in utility boardrooms; power was becoming cheaper year to year, and costs were steady or declining. Even though prices for other consumer items were rising, prices for power continued to fall. Indeed, in the years following 1962, rate decreases were far more common than rate increases.

Careful readers may jump to the conclusion that this fact runs counter to our fundamental belief that past is prologue. After all, there seems little likelihood that the price we pay for electricity will ever fall again, especially not on a continuing basis. While that may be true, the financial problems experienced by utilities during that era are not unlike those we see today. The operational exuberance that utilities exhibited during this period covered a multitude of sins—the most critical of which was a shaky financial foundation.

The utilities almost always claim that this perilous financial condition was attributable to the regulatory agencies' forcing them to act in the interest of short-term consumer benefit rather than in the long-term interest of the utilities themselves. Nothing could be further from the truth. The utilities spent money with wild abandon in the years following World War II, convinced that the investment would be repaid many times over by consumers eager for greater and greater amounts of electricity. What the utilities underestimated was the demand itself; their reserve power allocations did not reach satisfactory levels until the late 1950s. By then, spending was tapering off. The need for power plants had been met during the expansion years.

The point is this: Returns on utility investment actually rose during the period leading up to 1965. High internal overhead (the huge corporate CEO salaries we see today had their beginnings in investor-owned utilities in the 1960s) and increasing interest on poorly accounted long-term debt, on the other hand, contributed to the shaky financial condition that manifested itself in the late 1960s. As smaller utilities merged, and the big holding companies grew even bigger, the rather carefully concealed financial problems grew exponentially. The bubble was bound to burst, and burst it did—in a spectacular way.

The Great Northeastern Blackout

During the years from 1965 to 1970, Americans continued to increase their use of electricity. The price of power remained relatively flat compared to that of other energy sources, the overall cost of living, and a steep upturn in the price of coal, which still fueled most generating facilities.

Investor-owned utilities continued to grab market share, and they built bigger and more-expensive generating facilities that just did not work as well as those of their predecessors. More fuel was required to run the new facilities, and they stayed in operation for only part of the year. Nevertheless, stock prices for electric utilities reached their zenith in 1965. As the Vietnam War began to escalate, the real problem inherent in utility capital spending was about to reveal itself.

At 5:27 p.m. on November 9, 1965, the entire Northeast and a large part of Canada went dark. From Buffalo to the eastern border of New Hampshire, and from New York City to Ontario, a massive power outage struck without warning.

Trains were stuck between subway stops; people were trapped in elevators; failed traffic signals effectively halted the flow of traffic.

Amid the tension of the Cold War, many thought Armageddon had arrived. One pilot flying over a darkened New York City stated, "I thought, *'Another Pearl Harbor!'*"

By 5:40 p.m., 80,000 square miles of the northeastern United States and Ontario, Canada, were without power. More than 30 million people had literally been cast into utter darkness.

New York City was hardest hit by this blackout, largely because of the city's reliance on electricity for nearly all aspects of urban life. Office workers heading home for an evening with their families in the suburbs were forced to find alternative lodging. Some sought shelter in their offices or on the chilly benches at Grand Central Station. Theaters closed for the night. Times Square, usually a glimmering crossroads of light, was dark. Thousands of travelers stranded in New York were forced to sleep in hotel lobbies. The *New York Times* reported that "the city's hotels looked like bivouac areas." Nearly ten thousand commuters were stuck on subway cars, unable to escape the darkened subway tunnels. About midnight, the Transit Authority began sending food and coffee to those trapped underground. By then, many subway passengers had been stranded in the dark, without food or water, for more than six hours.

Despite the confusion and disarray, New Yorkers spent the evening of November 9 in relative peace. Just as in the hours following the World Trade Center disaster on September 11, 2001, there were no riots or widespread looting. Instead, New Yorkers helped each other. Some directed traffic through dark intersections. Others assisted the New York Fire Department, helping to rescue subway passengers. "The real problem," one rescuer commented, "was the fact that even the subway car doors were electrically operated. We were getting people out through the emergency doors—one at a time."

By 11:00 that night, the power had been restored to three-quarters of Brooklyn, and by 2:00 the next morning, electric power was back on throughout the borough. By midnight, much of the Bronx and Queens had electricity again, and at 6:58 a.m., almost fourteen hours after the massive blackout struck New York, power was restored citywide.

Six days were required to locate the cause of the blackout. Federal Power Commission investigators found a single faulty relay at Sir Adam Beck Station in the Ontario Hydro facility that had caused a key transmission line to open, or disconnect. This small failure triggered a sequence of escalating line overloads that raced through the main trunk lines of the Northeast's power grid, separating major generation sources from load centers and weakening the entire system with each subsequent separation.

As town after town went dark throughout southeastern Canada and the northeastern United States, power plants in the New York City area automatically shut themselves off to prevent the surging grid from overloading their turbines. Within a quarter of an hour, the entire grid was down. Investigators referred to the 1965 blackout as a "cascade effect," akin to a row of dominoes falling one after another.

Soul-searching

In the Federal Power Commission's final report on the Great Northeastern Blackout, the panel noted that the initial reaction to the power failure "was one of general disbelief that such an incident could happen."

Once the disbelief had faded, utilities and consumers alike were able to learn from the experience. Indeed, the 1965 Great Northeastern Blackout had ramifications for both power companies and electricity users. Initially the blackout forced many Americans to reconsider their growing dependence on electricity, and it prompted electrical engineers to reexamine the power grid system that had failed so unexpectedly.

Secondarily, the electric utility industry began learning how to plan for that sort of unexpected event. Regional coordinating councils such as the Northeast Reliability Council (NERC) and power pools like the New York Power Pool (NYPP) were formed to develop industry standards for equipment testing and reserve generation capacity, as well as to develop preventative measures governing interconnection and reliability, in order to preclude the possibility that a similar failure might happen again.

On a more personal level, New Yorkers learned to keep caches of candles, batteries, flashlights, and transistor radios in case of emergency.

Interestingly, obstetricians and gynecologists did a land-office business in the months following the 1965 holiday season. Indeed, August 9, 1966, exactly nine months after the blackout, marked the largest number of recorded births in a single day in New York City history. Clearly, some folks had found the prolonged power outage useful!

The blackout produced a psychological change as well. For the first time, both producers and consumers of electricity felt vulnerable. The memory of the blackout proved a disquieting reminder—utility officials and consumers alike could no longer consider their dependence or reliance on electrical power without thinking about the night of November 9, 1965. That is why, four decades later, the blackout still holds particular significance for those who lived through "the night the lights went out." One woman who spent her evening in a Lexington Avenue diner said, "This is the type of day when you remember everything—everything you did, everything you ate. You remember it all."

Quite clearly, the era of consumer innocence had ended—not with a bang, but with a flicker.

One might conclude that the nation's electric utilities absorbed the lessons of November 1965 a little too well. Electric companies began spending massive amounts of money in an effort to shore up service reliability. Unfortunately, this investment did not bring with it the same lower costs as had building power plants a decade before. Shareholder value dropped as revenues fell flat.

The power companies began spending money for other things as well, including, as a grudging acquiescence to the blossoming environmental movement, an effort to beautify their generating plants. This did not mean that utilities planted trees and shrubs on the generator grounds; instead, the power companies began converting generators to use more environmentally friendly fuels, although that term had yet to come into widespread use.

Now, it strikes us that anything done to improve a generating plant actually contributes to utility revenue in the long run. Unfortunately, that view is not shared by the utilities, which call such improvements "non–revenue producing." We side with the many analysts who believe that all capital improvements—even in pollution-control equipment, for example—are revenue producing. In any case, the cost for those improvements is almost always paid for by the utility's rate base.

The Embargo and Beyond

Suffice it to say that much utility spending during this period was independent of demand or need for additional electricity; hence, to the utilities it was non–revenue producing. The power companies bore their misfortune as best they could and bided their time. Strangely enough, the Yom Kippur War gave them the chance to strike back and to reclaim what they had always viewed as misspent capital.

In the wake of the conflict, oil-producing nations in the Middle East cut off shipments to the United States. Meanwhile, the OPEC cartel multiplied oil prices several times. Americans responded in typical—well, American—fashion by reducing their consumption of electricity. In large part, they did so because higher fuel prices were being passed along to the ratepayer through a fuel adjustment loophole in most regulatory agreements.

The higher price severely depressed the demand for electricity, but the power companies, which were selling at a higher price and passing all the increased costs along to the ratepayers, could not have cared less. Decreased demand, it was thought at the time, would give the utilities the opportunity to scale back their capital expenditure programs. The greed of the utility owners and managers, however, ensured that construction continued unabated. Utility officials thought the so-called energy crisis would sort itself out in due time. All they had to do was wait it out and then—wham! Ratepayers would be used to paying higher rates for electricity, which would, by then, be cheaper to generate.

Of course, the energy crisis was never really sorted out; it remains problematic to this day. Electricity usage never rebounded, and utilities that continued investing in new plant construction found those plants almost idle. Paying for them would be a financial burden over the next several decades. No matter. The same consumer who, by now, was used to paying higher and higher prices for a commodity once considered cheap and plentiful would pay for the idle plants as well.

In essence, the era of rapid electricity sales growth came to a screeching halt with the Arab oil embargo. The utilities' reaction was typically self-centered and inaccurate. As a result, consumers across the nation have, for decades, been paying for new power plants that have proved to be unneeded or, at the very least, underutilized.

Capital expenditures finally began to decline in 1975 but not because utility officials finally saw the light. They declined in large measure because the nation's biggest power company, Consolidated Edison, canceled its April 1974 common stock dividend. Investors had always taken for granted the notion that their investment return was a sure thing, but the Con Ed omission of the dividend shattered that notion. In April 1974 the price of the average utility stock fell nearly 20 percent. By 1975 utility managers were beginning to realize the sobering truth: They could no longer raise money at will.

The lack of readily available investor cash actually ended the utility expansion run. Faced with the prospect of paying market interest on borrowed funds, executives chose to curtail everything but their own paychecks.

By the late 1970s, the lack of utility attention to anything costing money was becoming evident. The nuclear accident at the Three Mile Island plant in March 1979 was emblematic. This was the United States' first major, well-publicized civilian nuclear mishap, which, we realize, is tantamount to calling a major airline disaster a navigation error. Residents of the Northeast were scared out of their collective wits by news reports of core meltdown, escape of radiation, and the potential explosion that might have ensued.

The Three Mile Island reactor finally went off-line, having incurred several hundred million dollars in damage. This single event destroyed the utility industry's rather complacent attitude about nuclear power, as the resulting antinuclear demonstrations attracted thousands upon thousands. Building a nuclear power plant became an unacceptable financial risk. If the plant went down, purchasing power from other plants to fill customer demand quickly became an expensive proposition. Worse yet, if regulators chose not to allow the utility to pass on the increased cost to the consumers (a rarity, but it did happen a few times), the utility might actually suffer a financial loss.

Such an event was too horrific for utility executives to contemplate, but it happened. General Public Utilities, subsidiaries of which had owned the Three Mile Island plant, was unable to pay investors a dividend or even to place securities in the public market following the accident. Investors who had embraced nuclear technology with the passion of hungry hound dogs suddenly turned skittish. The dream of nuclear power—to which utilities had subscribed largely because it promised such inexpensive generation that meters might no longer be required—vanished almost overnight.

Industry focus shifted from construction of new nuclear power plants to finishing those already begun. For some utilities, the financial strain of constructing the new facilities was too much to bear, and their financial standing was destroyed. A few utilities had never jumped on the nuclear power bandwagon for one reason or another; despite their prudence, however, they also faced an uncertain future.

The Modern Era

The 1980s was a time of readjustment for electric utilities, as they staggered beneath the financial burden of incomplete or unused construction projects, were battered by uncertain investors and high interest on borrowed money, and were besieged by regulators who were themselves besieged by angry consumers. There arose a fundamental concern that utilities might not survive.

They did, of course. Overall, the industry managed to overcome its problems. Utilities emerged from the dark times following the salad years to face a new reality: competition. For many companies, however, the demon of a competitive marketplace was nothing compared to the fiscal demons they had already conquered.

Beginning in the late 1970s and extending through the early 1980s, utilities found themselves at the mercy of the moneylenders. Interest rates soared during the latter months of the Carter administration and the early days of Ronald Reagan's presidency; at one point, every $1 million borrowed cost $250 in interest *per day*. The interest, of course, was being paid on money borrowed to finance or finish the construction of power plants, which, thanks to decreased demand, were largely unneeded.

That dark time passed unmourned by all except the financial institutions. The utilities were still in the deep rapids. In June 1983 the Washington Public Power Supply System canceled two nuclear power units and halted construction on a third. Investors sued the public power agencies that had backed WPPSS bonds via power purchase agreements. The Washington Supreme Court ruled that the agencies had never had the authority to sign contracts with the WPPSS in the first place. The money, of course, was long gone.

Four months later, Cincinnati Gas and Electric announced that its Zimmer nuclear facility would require another $3 billion and three years of work before it was finished. Investors, who had been under the utility's carefully crafted impression that the unit was nearly finished, were understandably shocked.

In the spring of 1986 the Russian nuclear reactor at Chernobyl went out of control. This accident, which involved the loss of thirty-one lives, and the resulting radioactive cloud over Europe doomed any effort aimed at a resurgence of nuclear power in the United States.

Less than two years later, Public Service of New Hampshire, staggering beneath the burden of building a large nuclear facility and hampered by consumer-friendly regulators, went bankrupt. From Washington State through the heartland and into New England, utility investors were learning a fundamental truth: They could indeed be wiped out by the utilities, and there was not much they could do about it.

By the end of the 1980s, consumers, investors, and government officials were becoming increasingly dissatisfied with the performance of electric utilities. Would competition do much to force more efficient performance? After all, the Iron Curtain had fallen, and free-market concepts reigned supreme the world over. So why not apply those same ideals to the electric utility monopoly? Finally, after a good deal of debate and a good deal of pressure brought to bear, the Congress acted. The Energy Policy Act of 1992 paved the way for competition within the electric supply industry. The act allowed more companies to become involved in the generation of electricity, and it freed up transmission apparatus so that new suppliers could get the power that they generated to the customers.

The Energy Policy Act, in essence, made power generation a viable business by lifting the restrictions that had been in place since the passage of the 1935 Public Utility Holding Company Act. The impact of potential competition took about a year to filter through to the investor base. The next Halloween, when the Edison Electric Institute held its annual financial forum, speakers declared that the competitive era had arrived and that utilities unprepared for the rigors of competition were in for a tough time.

When interest rates rose shortly thereafter, the market prices of utility stocks fell. As most analysts would tell you, this kind of nosedive reaction is usually the case.

The utility industry, however, portrayed its poor financial performance as a sign that investors were worried about competition. If investors were worried, the spin went, utility management must be in a blind panic.

Nothing could have been further from the truth. The advent of competition was an ingenious ruse. Gone in an instant were the age-old criticisms of electric utilities. Their self-serving nature, monopolistic tendencies, and so on were all buried under the notion that actual competition would finally force them to behave in a more civilized, businesslike manner. The real story, as we will see, is that regulators have failed at their task, and for the most part, competition has not really taken hold in the marketplace. As a result, electric utilities have run amok, unchecked and unchallenged. Few, if any, of the new competing firms have become large enough to make much of a dent in established utility business. Overall, the ratepayer base remains depressingly tied to the same utilities that have served consumer interest so badly for so very long.

The Art of Utility Empowerment

Individual states have enabled poor regulatory performance and actively promoted utility malfeasance. In a very real sense, states, and in some cases municipalities, have coddled individual utilities, providing a guarantee of revenue and profit, protection against loss, and even compensation for mistakes made in the course of expanding the utility itself.

A case in point: Gulf States Utilities. GSU made one of the classic business blunders that many utilities made during the 1970s: It committed too much money to build an unneeded generating facility. In GSU's case, the size of the mistake was so great as to force a normal business into receivership, but GSU was no normal business. Instead of the utility's going under, GSU customers were forced to pay exorbitant rates for electricity so that GSU could recover the misspent funds.

GSU's excuse was that the price of oil had unexpectedly increased, thereby decreasing the development of local industry and thus decreasing the demand for electricity. Do bear in mind that the area served by Gulf States Utilities was one of the nation's richest in terms of energy resources. Still, consumers there were forced to pay—and pay heavily—for an energy utility's cash mismanagement.

Incidentally, GSU's CEO was able to weather the storm in admirable style. Despite the company's troubling financial condition, he managed to retain his two jet aircraft and four pilots.

In any other industry—banking and real estate come to mind—investors would have been forced to bear the burden of the consequences of rising oil prices. State and local governments, however, saw no reason why GSU should be subjected to the same business standards applied to every other enterprise.

Spoiled they may be, but the utilities are not stupid. Because of their awareness of how the regulatory process favors them, they have learned to delay and drag out regulatory proceedings that might have an adverse impact on their business. This tactic also discourages utility customers from pursuing any sort of legal agenda against the power company; the system is rigged so that unless a customer has a claim worth $100,000 or so (and few do), no economic gain is realized by seeking relief through the regulatory process.

When litigation does proceed—when large, economically viable claims do actually go to court—and the utility is found liable, the ratepayers' pocketbooks protect the utility from real financial harm.

Until the era of deregulation, utilities regularly utilized the media outlets to advertise their splendid service and exemplary employees. Ad campaigns were designed to make people feel good about writing a check to their local utility company each month. Since deregulation, of course, the tone of the advertising has fundamentally changed; the utility now stresses a long record of good service and forcefully suggests that customers avoid unnecessary hardship and, although difficult to imagine, potentially higher rates by remaining loyal.

A Business Like Any Other?

The theory behind the regulatory structure was to protect the consumer against unreasonable power and force that might be wielded by a monopolistic utility. The regulatory body was supposed to simulate the forces inherent in a competitive marketplace. Bureaucracy seldom imitates real work; that is why the regulatory process has, in general, failed miserably. For much of the era following the salad years, regulators in state after state allowed utilities to earn back their capital expenditures. More recently, the problem has been with how those assets have been managed and manipulated. We will discuss this point in greater detail later in this book, but suffice it to say that utilities have consistently valued their fixed assets at rates that far exceed the market value. As a result, consumers footing the bill have paid for those assets many times over—and continue to do so. Additionally, investors and financial institutions have been misled by these excessive asset valuations . . . but that's a different story.

Herein lies the real paradox: Utilities always portray themselves as just another business, just like your neighborhood pharmacy or grocery store. The truth is, utilities have no real similarity to any other sort of free enterprise. Through the process of regulation and, in modern times, because of the lack of real competition, utilities are a breed apart—something akin to the bastard child of free enterprise.

Unlike other businesses, utilities have been shielded from the uncertainties of the marketplace, protected against natural and man-made disasters, and held harmless in the case of malfeasance or ineptitude. In college and graduate school programs, prospective entrepreneurs do not study the way in which utilities are managed precisely because bankruptcy often results when that standard of operation is emulated.

Regulatory agencies were established to govern utilities amid the seeming absence of change, and yet change is a constant; the world around us changes with each passing hour. While we sometimes tend to think that the pace of change has accelerated, we must not overlook the fact that even during relatively tranquil times change has been a constant.

As things change, technology inevitably advances. Yet regulators tried for decades to shield utilities against the market forces brought to bear by new technology. In the electricity marketplace, in particular, the advances of technology served to spur growth, not to provide competition from new energy sources.

Here is one way to think about utility operation in the midst of change and advancing technology: As the expense required to produce the product decreases, the cost of that product naturally declines as well. For our nation's electric utilities, the era of big power plant development faded away decades ago. The plants that were built during the salad years to generate the electricity you use today have long since been paid for and depreciated. Therefore, the cost of the electricity those plants produce should decrease as well.

However, most people have seen only a modest decrease in their monthly utility bills, and a disproportionate share of the rate-paying population has actually seen their bills increase.

Your monthly electric bill is unlikely to decrease in the future, even with the advent of competition and a deregulated marketplace. As we shall see, the vestiges of regulation are, in effect, strangling the effect of competition among utilities. Despite what the utilities claim, the playing field is still far from level. As long as the company that owns the wires that connect to your home or business has a monopoly, the playing field will probably never be level.

Accountable Only to Themselves

We live in an era of labels. In politics, liberals believe conservatives are dogmatic and inflexible. Conservatives believe liberals have one set of rules that they apply to themselves and another set that they apply to everyone else. Well, on the basis of popular ideology, you would have to classify electric utilities as extreme liberals, since the normal standards you and I have to follow to survive in this world just do not seem to resonate with utilities. For whatever reason, notions of accountability and proper business behavior seem to have little applicability in the world the power companies inhabit.

For example, the concepts of value-based customer service and fair and equitable treatment got tossed out the window in Houston some years ago, when Houston Lighting and Power sent a letter to a local hospital claiming that its test of an electric meter at the hospital showed that the meter was inaccurate. According to the utility, the hospital had used, but not paid for, a certain amount of electricity; the utility backbilled the hospital for six months' worth of power that it said the hospital had used but that had not been metered.

As we examined the hospital's billing records for the previous four years, we found that the highest consumption period had occurred during the same six-month period in question.

Our office sent a letter to the utility explaining that the situation was not equitable and requesting a rebill. All we received in reply was a terse note. According to the utility, the billing was accurate. A second letter elicited the same result.

Finally, we were able to have a telephone conversation with a utility executive. "When we find a meter is inaccurate," we were told, "we just routinely backbill for six months. We don't check to see how long the meter might have malfunctioned . . . . We just backbill all the errors for six months' usage. It kind of evens out over time."

"That may be true," we responded, "and it may be a convenient way to balance your account ledgers, but you've overcharged our client more than $11,000."

"W-e-l-l," said the executive, "that's just the way we do it. If you don't like it, why, that's just too bad." Then he hung up.

The message was clear enough: The power company really did not care how its errors affected everyone else; all that mattered was that accounts averaged out for the utility. We finally had to threaten to discuss the $11,000 discrepancy with state regulators. Faced with having to defend its position in a public forum, Houston Lighting and Power eventually refunded the hospital's money.

Doubtless there are other businesses in which an executive could get away with telling a customer, "If you don't like it, that's just too bad." At the moment, however, none come to mind. When we are asked if utilities are responsive, friendly, and accountable, we always respond in the affirmative. Power companies are accountable—to themselves alone.

A similar case took place in Mississippi. A shopping mall had been billed at the wrong rate. We sent a letter to the power company, asking only that the utility change the rate and refund the overcharges, which totaled about $20,000.

After several weeks of delay, the utility wrote back, agreeing to change the mall's rate but refusing to refund the overcharge: "It is against our policy to make refunds," the letter stated.

Finally, after another round of letters, we threatened to sue the utility for the $20,000. The utility sent my client a check for just under $12,000. As we pondered the rather peculiar figure, it dawned on us: The electric company knew we would spend just over $8,000 to take the case to court. The refund was the least amount the utility thought we would accept.

Going to court for $8,000 was prohibitive, since the case would consume far more than that amount in time and expense, but we continued to contact the suddenly tight-lipped utility and even took the matter to Mississippi's Public Service Commission.

That got their attention—not that the utility executives were particularly worried about what the PSC might do, though. Instead, the utility's general manager wrote to the manager of the mall, saying, "This high-pressure tactic is, in my opinion, not the way solutions to problems are to be found." The letter made us wonder if the utility executives thought pistols at dawn might be more appropriate.

If you choose not to communicate with someone and refuse to own up to your honest mistakes, do you have a right to expect others to believe you are accountable for what you do? If you think about it, you will see how the popular definition of the word "liberal" applies to the utilities. Apparently, a different set of rules exists for their customers than the ones that apply to them.

Eventually, the utility made off with the $8,000, and our client never got the refund. By calculating down to the penny the amount of money it had to return, the power company effectively stole thousands of dollars from our client, which, of course, had no choice but to continue to do business with the same sorry folks.

If your local utility cannot steal money outright, it can often borrow money for free without bothering to wait for a favorable economic climate. How? Just overcharge customers and wait for them to figure out they have been overbilled. Meanwhile, the utility often has free use of customers' money, since utilities in many states are not required by law to pay interest on overcharges.

Before 1987, when we enjoined a client to successfully challenge the law, utilities in Texas would not pay interest on overcharges. As the law stood, Texas utilities could overbill someone hundreds of thousands of dollars, hold on to that money for years and years, and pay no interest if and when the customer finally discovered the error.

Indeed, the concept of money as an earning vehicle is apparently lost on electric companies, unless the utility is lending its own money. Then, of course, it would expect to be paid interest.

This absurd double standard stood unchallenged in Texas until one of our clients, the Dallas County Youth Village, decided to take on TU Electric in an effort to change the status quo. During our audit, we discovered that TU Electric—now known as TXU—owed the Youth Village a refund and about $300 in interest. For that $300, we fought the case all the way to the Texas Public Utility Commission.

First, we had to convince the Dallas County commissioners to let us pursue this case as a legal test. County judge Lee Jackson and several commissioners bucked considerable political pressure from TXU and authorized us to move forward. Eventually, we succeeded in getting the PUC's rules amended so that utilities were required to pay interest on overcharge refunds.

TXU fought us every step of the way. In its comments before the PUC, TXU maintained that interest payments should not be required because overcharges were not deliberate and resulted only from "purely unintentional human error." What TXU wanted, in essence, was preordained immunity for overcharging customers. This desire, in and of itself, is a good example of how the monopolistic mind-set has steadily corrupted the soul of the utility. We believe that had TU won the argument, the utility would have happily yielded to the officially sanctioned temptation to overcharge customers.

If you make a mistake involving other people's money, your creditors are liable to require interest and penalties nonetheless. In the real world, it really does not matter if the mistake was caused by "purely unintentional human error" or if you meant to make the error. You would owe the interest just the same. Electric utilities apparently are not bound by the same considerations. TXU maintained that if the PUC decided to require the utility to pay interest on overcharges, then the utility should be allowed to collect interest from customers who had been undercharged.

TXU attorneys noted:

> The *only* legitimate function or purpose for imposing any interest requirement is because one party has had the use of another party's money for a relatively short period of time. Thus, simple notions of fair play dictate that, if there is to be any interest requirement at all . . . *then interest (at the same rate) should be imposed in cases of underbillings* (emphasis ours).

We invite you to briefly consider the difficulty inherent in imposing this odd double standard. Can you imagine telling clients you have *underbilled* them and that, despite your mistake, they are required to pay interest on the money they owe you? Try this with an independent businessperson and see how fast he or she gives you the boot!

The utility made it quite clear that if the commission forced it to pay interest on overbillings, the cost of that interest would be included in the utility's cost-of-service figures for rate-making purposes. Eventually, TU Electric customers, not the utility, would pay the interest.

As the Public Utilities Commission announced the ruling that utilities must pay interest on overbillings, it made note of the fact that our case involved only $300 in interest and that each side had spent far more than $300 to bring the case to the commission. "The evident hole in the rules," as the chairman called it, had effectively been plugged. Although TXU and other utilities would put up a big fight to avoid being domiciled in Texas, they now had to play on the same field as every other business.

This is as it should be. We live in an era of accentuated accountability. We demand accountability from our government, our schools, our elected officials, and most businesses we patronize. Unfortunately, we have not typically demanded the same accountability from our utilities.

Why? Because to demand and expect some degree of accountability, we must first understand the inner workings of the system we scrutinize. Most of us have at least some rudimentary understanding of how government works—the basic principle of tax-and-spend is not hard to grasp. Most of us have been through the educational system in one format or another and thus can bring our own personal knowledge to bear when we demand accountability from our schools. We can demand accountability from businesses because we know something about the product we are purchasing. However, electricity, the system that delivers it, and the power utilities themselves are deep mysteries to most Americans.

For the electric utilities, consumer ignorance is bliss. Unburdened by real competition and disdainful of their own customer base, the power companies have been able to do pretty much whatever they want.

How Regulators Really Work

In both Texas and Mississippi, as was the case in most other states, regulators were nothing more than a thorn in the side of the power companies. The threat of regulatory attention or action was merely an inconvenience. The utilities' complacent attitude toward regulators is widespread, and with good reason: As we have already noted, few entities have done more to aid and abet power company misbehavior than the very regulators that were created, elected, selected, or otherwise empowered to guarantee that consumers would not be abused by their local electric monopolies.

For years, we have used the term "spoiled child" to describe utility company behavior. "A spoiled child, left alone," we have often said, "almost always gets into trouble." Similarly, regulatory agencies have frequently assumed the role of indulgent parent, overlooking or ignoring utility misdeeds until the acts encroach on public awareness. So long as what the utilities do is transparent or unseen, most regulatory bodies could not care less.

That is not to say that public service commissions and the like are not made up of good, decent citizens. They are. The problem is that, in general, regulators do not concern themselves with quality-of-service issues. If your electric provider acts like a spoiled child, the regulators are hardly likely to sit up and take notice.

So what is the real purpose of a regulatory body? Well, in an ideal world, the regulator would take the place of a competitor. Regulators are supposed to replace the competitive spirit found in most areas of enterprise with governmental imperatives designed to help utilities live up to the spirit and letter of their compact with the area they were appointed to serve.

Because the customer has no choice in a non-competitive environment, the regulatory agency is supposed to ensure that the utility delivers good service. However, regulators cannot, and typically do not, concern themselves with the quality of that service unless and until it degenerates into a degree of uncertainty untenable for most other businesses.

Instead, regulators spend their time studying the apportionment of costs of electric service in an effort to make sure that no group of customers is unfairly charged. Additionally, regulators set the level of utility revenue at a point high enough to allow the electric company to earn a rate of return similar to that earned in industries where competition exists.

This fact seems something of an entitlement to utilities, especially when they obviously intend to serve themselves at your expense. Much regulatory activity is aimed at one task: granting rate increases to utility applicants. These rate increases affect everyone from large corporations to millions of residential customers.

For regulated utilities, the typical residential customer was tantamount to a small flea on the back of a very large dog. Grouped together, however, all these fleas did do something that the utilities could enjoy and appreciate: They paid the utilities' costs of development and construction of new facilities. All too often, public service commissions actually came to the aid of the utilities by shifting this back-breaking financial burden onto the consumer. This practice had a negative effect upon everyone, but the bulk of the impact fell upon people who could least afford it: residential customers who had a tough time paying their utility bills.

As deregulation advances, however, regulators are more apt to pay attention to individual consumers, because those consumers represent votes that keep the elected officials who appointed the regulators in office.

In ancient times, those who owed a debt found themselves in servitude to their creditor. The utilities view ordinary consumers in much the same light. Somehow, they have come to believe that *we* owe *them*. As a consequence of this attitude, utilities frequently see the public as a servant of sorts—a beast of burden created to bear the loads the utilities do not want to carry by themselves.

In Texas, TU Electric asked the Texas Public Utility Commission for permission to add all of the cost of developing a nuclear power plant to the existing rate base. In simple language, this means that every TU customer must share in the burden of building a nuclear power facility. TU, of course, would generate power from this free source, charge customers for that electricity, and show a handsome profit to stockholders.

Prior to appearing before the PSC, TU executives admitted to some problems at their nuclear plant—problems that ended up costing lots of money. Some of the problems, the utility finally confessed, were its own fault. Yet TU had no compunction about asking consumers to pay for the mistakes that TU itself had made.

We have no problem with a utility's asking for partial reimbursement for large developmental expenses. We have a big problem, however, with consumers' being expected to foot the entire bill, especially when the utility is making half a billion dollars a year in profit! This, incidentally, is the principal reason why utility stocks have always been considered such safe investments. Customers, not stockholders, pay for management mistakes!

Portions, Prices, and Profits

If you choose to buy a McDonald's restaurant franchise, you invest a fair amount of money and are secure in the knowledge that McDonald's will not sell another franchise just across the street; in other words, the franchisor will not compete with you, at least not directly. Power companies shelter themselves from competition beneath the "golden arches" of their regulatory agencies. The utilities get what McDonald's cannot afford to give: an exclusive franchise, period.

Like an eager politician gerrymandering a local district to secure a given block of votes, utility regulators have defined the service area for the power companies. Block by block, subdivision by subdivision, regulators have assigned the task of distributing and marketing electricity to already established utilities. The process is typically not an intricate one; most often, only one utility serves a given geographic area. Consequently, there is usually very little apportionment to be done.

In the past, when two rival utilities competed for the same customer base, the regulatory body often found itself in a no-win situation, something like McDonald's might face with competing franchisees a block or so apart. Pleasing one utility typically meant angering another, and the power company that won the battle often had exerted undue influence over regulators. Many of the mergers and acquisitions common in recent utility history owe their origin to regulatory efforts to divide a marketplace equally among rival power companies.

Fortunately for regulators nationwide, territorial squabbles were relatively uncommon. Regulatory agencies typically concerned themselves with slightly more mundane matters: setting the price for electricity, determining the allowable utility profit margin, and monitoring the cost of the product the utilities sold.

Since there was money to be made through the sale of equipment rather than the sale of electricity, the early years of the electric industry were dominated by manufacturers. Once monopolistic practice gained a foothold, the utility industry invested heavily in generation facilities, investments that continued through the salad years and beyond.

Today, however, rate-based investing is somewhat archaic. Stockholders and investors do not want utilities to invest heavily because those expenditures have been handled so badly in decades past. Therefore regulators face a new and different challenge. If the old ratios determined by dividing revenue by capital expenditure no longer hold true, how will rates be determined?

While prices at McDonald's remain relatively consistent from coast to coast, prices for electricity vary widely. For example, consumers in Kansas pay vastly different rates for power than do customers in Maine or California. In regulatory practice, one could hardly expect otherwise; different locales depend on electricity generated by different types of plants, which themselves require varied investments.

If those investments have long since ceased and are no longer a part of the equation, what keeps power companies from overcharging their rate base to engender an ever-increasing profit for themselves? Almost nothing.

Deregulation was supposed to introduce competition to a non-competitive environment, thus suppressing prices by the power of choice. The reality, however, is that few competitors nationwide have managed to gain more than a toehold in their respective marketplaces. As a result, competition has not yet proven to be a particularly effective way to manage the marketplace.

Regulation is not without fault. We freely admit that regulated monopolies, by their very nature, tend to create more problems than they solve. For example, you do not have to be a particularly astute observer to conclude that had regulators been doing their jobs, prices for electricity at the consumer level would have fallen over the past few years as the utilities paid off the excesses and investments of the post-1965 era.

The old investment-versus-return formulas just do not seem to work in the modern era. Unfortunately, competition as a substitute has proven a miserable failure. As we will see in the next chapter, electric utilities have taken advantage of the regulatory lull in imaginative and creative ways—all of which are designed to enrich the utilities at the expense of the customer.

2

THE REWARDS OF INCOMPETENCE

The Blame Game

Four decades have passed since the Great Northeastern Blackout in November 1965—four decades that have seen the promises of utility service improvement and the inherent benefits of deregulation lying fallow at our feet. Or, if you are still charitably inclined toward electric utilities, you might conclude that God just has it in for New York City.

The blackout of 1965 affected 30 million utility customers and covered 80,000 square miles. Many of those same folks were hit again—this time in the midst of the staggering July heat—in 1977. After lightning struck upstate power lines, 9 million New Yorkers were without power for more than a day. Nearly 4,000 of the more enterprising city dwellers were arrested for looting.

Nine years later, a substation switch malfunctioned and blacked out a four-block area of the Big Apple for twelve hours. Then, two days after the Fourth of July in 1999, record-breaking heat caused power lines to expand and arc, blacking out the city for nineteen hours.

This most recent New York blackout was, in the words of a frustrated utility engineer, clearly God's fault. "Expansion and contraction are fundamental physical principles," the engineer explained. "All the utility had to do was to find a way around God, and the entire blackout could have been avoided."

The real problem, of course, was that delivery equipment hadn't received the attention—repetitive, boring, and costly though it was—that the lines deserved and demanded. Checking the lines was nothing but an expense to the utility; by avoiding the issue altogether (and blaming the results on God), the power company escaped with its integrity and reliability of service reasonably intact. As a result, the utility bought itself four years' worth of time, arguably at God's expense. When God finally caught up with it, He did so with a vengeance.

When the lights went out on August 14, 2003, the blackout crippled the northeastern United States and parts of Canada. Ultimately, this outage was blamed on a software glitch; apparently an engineer who had disabled a software alarm "trigger" went to lunch without resetting the alarm. Two server overloads later, the entire region went dark.

Within weeks of the 2003 blackout, energy industry spokespeople were calling for grid investments of nearly $60 billion, along with as much as $450 billion in other utility infrastructure improvements. Spending massive amounts of money on new wires, power plants, and conventional fuel sources, the industry believed, would solve the reliability problem.

Administration and congressional leaders could hardly blame God for this disaster. Instead, they jumped on the utility improvement bandwagon, demanding higher profits for transmission line owners (apparently believing that some trickle-down economic effect would cascade higher profits into solutions to the reliability problem), as well as federal eminent domain powers over new transmission lines. All this public utility pork was rolled into an overall energy bill that included tens of billions of dollars in additional utility incentives and a rollback of certain consumer protection legislation.

Indeed, within a day after electrical service was restored to customers in New York, Cleveland, Detroit, Toronto, and other areas along the East Coast and throughout the Midwest of the United States and Canada, the secretary of energy said, "The people who benefit from the system have to be a part of the solution here."

Ratepayers were again being told to pony up. The only surprise in the secretary's comment was that God wasn't cited as the source of the blackout.

"In announcing that rate increases will be imposed to pay for the upgrade of the electrical transmission system," one rather clear-thinking newspaper editor noted, "the administration is repeating a familiar pattern: Policies pursued in the interests of an elite section of the population have created a social disaster—this time in the form of a blackout that affected 50 million people. But it is the ordinary people who will pay for the disaster . . . . Instead of compensation, the government promises only higher costs."

The utilities fought this public relations onslaught the only way they could. The power companies tried to shift the blame for the blackout on burgeoning deregulation. The industry correctly pointed to the fact that electricity generation represented only 2 percent to 3 percent of the American economy, while the other 97 percent depended on how well the first 2 percent to 3 percent worked. The true best interest of society, the utilities argued, does not center around the lowest possible electric rate. Instead, that interest is best served by a completely reliable source of electricity produced at a reasonable price.

Besides, the industry argued, whatever benefits that might have been derived from deregulation had already been spoiled by the likes of Enron—greedy, dishonest energy marketers who had ostensibly ruined the potential of deregulation by stacking the deck against consumers from the start.

According to the utilities, the old regulatory system—the system that had allowed the power companies to flourish for so long—was also to blame for rate hikes. The utilities maintained that regulators had been too slow to approve construction of new power plants, thus creating an electricity shortage. This was a hard contention to sell, given the heady expansion days of the 1960s and 1970s. Virtually ignored were hundreds of still usable but redundant power plants (known to the utilities as "stranded assets") dotting the American landscape.

Undaunted, the utilities claimed that regulators had insulated retail consumers from rate increases even while the wholesale costs of electricity rose. The fact that wholesale costs rose to pay for things like corporate jets and exorbitant executive salaries wasn't mentioned. Finally, the utilities maintained that major providers, forced by the states to buy power from independent electricity producers, could not sign long-term contracts and thus were committed to pay prices that rose on a daily basis. After all, how could utilities really be expected to face the same constraints as any other business?

Regulation, the utilities argued, was simply a veil that effectively covered up the bad choices of the regulators while forcing the utilities into an increasingly untenable position. Deregulation had been an even worse experience, in that the lack of governmental shielding kept profits at a minimum for the sake of competition—competition that the utilities neither needed nor wanted. Clearly, the only entities held blameless in this delivery debacle were the electric companies themselves. According to their side of the story, they had done everything right.

The Problems with Regulation

Clearly, we believe that the entire concept of utility regulation is inherently flawed. Much like a trick two-headed coin, regulation is designed to guarantee a favorable outcome regardless of the circumstance or situation. Unfortunately, things seldom work out that way. In recent years utilities have received an unfair share of regulatory privilege while consumers have gotten the shaft.

Regulators are appointed to act in the public interest, as public trustees, safeguarding consumers from utility excesses. But they are also appointed to ensure that the power company delivers quality service and is able to generate an acceptable profit. Initially, regulators and local and state governments didn't seem particularly bothered by the incompatibility inherent in those two objectives. That rather tranquil time passed rapidly away as utilities learned to manipulate the regulators.

Imagine a regulatory body happily going about the business of setting rates and ensuring a fair profit for the utility, which, in turn, is providing a needed service in a monopolistic environment with some degree of dependability. Then one day, probably very soon after this idyllic scenario first commences, the utility discovers a strategy that will ensure favorable regulatory treatment and increased profits—both at the same time.

"Dear Regulators," the plea might begin. "We don't have enough money. If we don't get more money, we'll probably go broke. And if we go broke, millions of consumers who depend upon us for one of life's basic necessities will have to do without.

"Since you are sworn to protect the interests of the consumers, you are thereby compelled to allow us to raise our rates. We will then be able to generate enough money to avoid going broke, and we can ensure that the consumers you safeguard will have enough electricity to satisfy their needs.

"It is, therefore, your sworn duty to allow us to raise our rates."

Indeed, the utilities have spent billions of dollars hiring attorneys, economists, and consultants and charging them with finding ever more creative ways to repeat this message. The electric companies understand that spaced repetition has an impact on the regulators they seek to influence. Put yourself in the place of a regulator and ask, How could you refuse? How could *anyone* refuse?

Conflicted from the Start

This is the conundrum of utility regulation: Regulators empowered to ensure consumer delivery of electricity are at the mercy of the utilities they regulate, and those same utilities have no compunction about threatening the regulators with dire consequences if their demands for additional revenue are not met. Such demands, of course, almost always run counter to the interests of the same consumers the regulators were empowered to serve and protect. And so it goes.

Some years ago, a small group of municipal governments brought us an interesting case. Their electric provider had been consistently overcharging these city governments; indeed, it became immediately apparent that the utility in question had violated the specific terms of the tariff approved by regulators.

We found the overbilling to be a widespread practice. If a ruling found the power company liable for the overcharge, the result would be a costly refund . . . very costly. To make matters worse from the utility's perspective, it appeared that the municipalities that had hired our firm had an open-and-shut case. At the outset, the attorney representing the utility advised his client to prepare to lose the case.

Fortunately for the utility, such cases are not heard before a judge and jury. Instead, they are tried before an administrative law judge, who acts as judge and jury. The administrative law judge, however, does not have the last word on the matter. His or her recommendation is passed on to the regulatory body, and a vote is held to determine the final outcome.

For this particular case, our administrative law judge was a young attorney. Suffice it to say that as a rule such judges do not find their position financially rewarding. We can imagine several better ways for enterprising young attorneys to make money.

To the surprise of almost everyone involved in the case, our administrative law judge strongly recommended to the regulatory body that it rule in behalf of the power company. Then, by a narrow vote, the regulators themselves decided to accept the judge's recommendation. But the story doesn't stop there.

Immediately after the regulators adopted his position, our administrative law judge resigned and went to work for a particular law firm. Exactly one year later—one year to the day, in fact—the young man made another job change and went to work representing the same utility before the same regulatory body.

A state law requires a waiting period of precisely one year before one can leave a position working for a regulatory agency and go to work representing a utility. Certainly coincidences are possible; unlikely events do occur. But this one hit a little closer to home than most. We were not particularly inclined to dismiss the incident as mere happenstance.

In another case we pursued through an administrative law judge and to a panel of regulators, we were surprised yet again. This time, in an open session, a regulator claimed that the utility in question was blameless. "I know these guys," the chairman of the regulatory panel announced to all and sundry. "They would not do this."

Yeah, right!

When administrative law judges proffer opinions to regulatory bodies, the regulatory staff itself often files motions in support of one side or the other. You'll remember mention of a Texas-based case we won that forced utilities to pay interest on overcharges. Well, in that particular case, we were dismayed to find that the regulatory staff had filed a petition siding with the power company, that power companies should not have to pay interest on overcharges.

Fortunately, the staff opinion did not hold sway. We eventually won when the issue went to the commissioners. But this is the best example we can cite to prove a single point: Regulatory agencies and those who staff them often serve as lapdogs for the utilities.

The utilities, for their part, have learned over the decades the best ways to cover their hammer with velvet. As we'll detail in the following chapter, huge amounts of utility capital and effort have been employed to become better "connected" with regulatory officials. Utility officials patronize regulators at civic clubs, in church, and by working alongside them in their favorite charity. To make matters worse, power companies routinely hire staffers away from the regulatory agencies; in fact, many regulatory staff members dream of a big job with a big future working for the power company they are supposed to be regulating! And, true to form, the utilities hire enough former regulatory staff members to justify keeping those dreams alive.

Here's the unsurprising bottom line: *Utilities make their money at the behest of legislators and regulators, not by virtue of the quality of service they provide to ordinary consumers like us.* If we were in the same situation, we too would be "schmoozing" our socks off! For better or worse, the success of utilities lies in the hands of those who are also charged with safeguarding the interests of consumers. For utilities to succeed beyond creating for themselves a modest profit, they must assist regulators in compromising those safeguards. Time and time again, the utilities find that the regulators themselves happily aid and abet the power company's best effort.

How Things Work

Today's utility ratepayer is probably more confused than ever before. Deregulation means something vastly different to consumers than to electric companies, although most of what consumers read and hear has been carefully crafted to represent the utilities' point of view. The standard line has utilities as champions, vitally interested in protecting our air, water, wildlife, and pocketbooks. In the last case in particular, nothing could be further from the truth.

Utilities would have us believe that they, along with their customers and regulators, are all involved in deciding how much to spend on programs that protect our environment, help needy families pay their energy bills, and ensure safe and reliable power service. As regulated monopolies that do not face direct competition, utilities have been given the long-term flexibility to invest in whatever they choose. Judging solely by mass media advertising bought and paid for with your hard-earned money, you would think that utilities are most concerned about energy conservation, solar, wind, and other renewable energy resources, low-income bill assistance, salmon protection, and other vital programs.

In the next two chapters, you will see that utilities are vitally concerned about only one thing: making money—or the nearest alternative, which is putting forth an appearance of making money. For the sake of argument, let us continue.

How have utilities traditionally determined rates?

Investor-owned utilities (IOUs) set electricity rates based on all the costs of providing electricity plus a reasonable profit. IOU rates are determined by state utility commissions. In return for a guaranteed base of customers, IOUs are obligated to serve every customer in a given service territory.

Consumer-owned utilities (COUs), on the other hand, operate as nonprofits and are obligated to serve all customers in their local market area. COU rates are set by local elected officials. When you hear a utility commercial claim, "We're here to serve you, not to profit from you," it is a cinch that the provider is a COU.

Rates vary according to customer class (homes, small businesses, or industries). For example, large-scale businesses and factories may be cheaper to serve because they buy very large quantities of power used at one location. These big businesses pay less per kilowatt for their electricity than average homeowners. The rate-setting process can be contentious; as you might expect, utilities, customers, regulators, and local officials do not always agree on what is fair for different customer classes. Unfortunately, the utilities' view almost always prevails.

Electricity rates are a combination of three costs: generation, transmission, and distribution.

First, let us consider the cost of generating the electricity or purchasing it from another utility or power producer. Most utilities get their electricity through some combination of owning and utilizing their own power plants and buying electricity from other generators.

The cost of producing power breaks down into basically two parts: first, paying off the long-term capital costs of building power plants and, second, annualizing the costs of maintenance and fuel (coal, natural gas, nuclear, etc.). Of course, the costs of building and running different types of power plants vary greatly, and the construction cost is the major factor that determines the different rates that utilities charge. Regulations guarantee that utilities have a way not only to pay for investments in power plants but also to generate a solid profit.

Next, one must factor in the cost of transmitting electricity from power plants to substations in cities and towns via high-voltage transmission lines.

Finally, distribution costs encompass the distribution of electricity to the individual homes and businesses where it is ultimately consumed. Typically, regulated utilities own their own distribution systems.

Now that you understand the costs involved in providing electricity, let us briefly discuss the current regulatory climate. First, just what is deregulation?

Under deregulation, also called restructuring, utilities would no longer have a guaranteed customer base or service territory. Utilities that both generate and distribute electricity would compete with each other and with companies that only generate power. Without monopolistic protection, the ideal situation is competition to provide customers with the best service.

In theory and in popular advertising campaigns, customers can choose which company will provide them with electricity. How utilities will pay for investments in power plants made under past regulations is uncertain. One thing, however, remains clear: The ratepayers, rather than the utilities, will bear this burden as well.

Rates can vary sharply, according to the market price of generating electricity and the associated costs factored into that price. Under deregulation, customers would no longer have guaranteed rates and would face the uncertainties inherent in a market-based energy pricing system.

The need to generate a profit, and the desire to eliminate debt from building nuclear and other power plants, ostensibly prevents many existing utilities from offering power at current low market rates. Low market prices combined with large-scale businesses pushing for access to that low-priced power provide the driving force behind deregulation. Many analysts believe that large businesses with market clout would benefit most from deregulation and that many residential and other small businesses would see little benefit or even a rise in rates.

Utilities, unaccustomed as they are to competition in any form, will argue that deregulation is already hurting our air, water, wildlife, and pocketbooks. The fact that the utilities themselves have been doing just that for decades is somehow lost in all the hand-wringing.

Faced with competition for the first time, utilities are already trying to lure large, desirable customers, such as industrials, with offers of cheap electric rates. Ordinary customers get the shaft; residential households will likely end up with higher rates as a result. To further cut costs, many utilities have chosen to slash programs that guarantee clean energy through conservation and renewable resources, such as wind and solar energy. Commitments to protect fish and wildlife and programs that help families in need to pay their bills are also at risk. God forbid the utilities should consider slashing their own considerable profits!

The utilities' message is clear: In order to restore funding for these vital programs, a protected playing field is essential. In other words, utilities want to establish a way in which they can generate greater profits and yet remain competitive. Many electric company ad campaigns make much of something they call a "system benefit charge," which is ostensibly designed to help ensure that important conservation programs are not swept aside in the upheaval of deregulation. Of course, profits would remain intact as well.

In an ideal world, how would deregulation work? Obviously, generation is the point that ensures a competitive market. Current market prices have largely been determined by the cost of new, more efficient natural gas combustion turbines. The initial construction costs for older power plants (mostly coal-burning plants) have been paid off.

For competition to work, the transmission system would have to be run independently of generation. In other words, no utility that generates power should control any transmission lines. Otherwise, utilities could unfairly discriminate against other power producers by overcharging them for transmission services. Transmission would remain a regulated service.

Additionally, distribution lines would remain a regulated service. Some utilities might elect to sell their power plants and transmission lines and become distribution companies. Deregulation of wholesale electricity has already begun under the Energy Policy Act of 1992.

One of the potential outcomes of deregulation is the formation of new companies called aggregators. (You may have heard other terms, such as "brokers," "retailers," and so on). Ideally, aggregators would serve to link customers, power generators, transmission operators, and distribution companies. Aggregators would then compete with each other for customers by offering the best generation price or customers' preferred generation source, thus appealing to the millions of Americans who are concerned about energy conservation and favor alternative methods of power generation. In the process, aggregators would also negotiate transmission and distribution services for customers. We hope that the aggregators will do a better job than the regulatory bodies they are designed to replace, and because aggregators will be competing with each other, we have reason to be optimistic.

At best, under the current regulatory system, utilities have little incentive to be efficient, effective, and responsive. Poor managers are bailed out by a system designed to protect incompetence; better managers and stewards earn no reward for their efforts. Cost reduction and increased efficiency might subject the utility or its holding company to substantial financial risk, for innovation almost always carries a steep price tag. If the efforts toward better performance fail, however, the company might be penalized by the regulators.

One of the arguments we will examine at the end of this book involves the question of socialization of public utilities. As entrepreneurs, we are naturally opposed to government ownership of public utilities, although we can certainly see some advantages to it. After all, utilities owned by the government would not be run on a profit basis—nothing else under government control is expected to make a profit. Regulation would therefore become unnecessary.

The few government-owned utilities in the United States today do operate without much interference from regulators at any level. Because investor-owned utilities ostensibly perform social services from time to time, one could argue that they have already been socialized—but that is far from true.

Electric utilities in the United States are, for the most part, static monopolies and will likely remain static monopolies until another technological innovation comes along to take their place. Those who argue that the concept of utility monopoly is outdated or no longer applicable ignore a fundamental truth: Utilities, by and large, continue to act as if consumers have no other recourse.

In our shoe-leather business, we have uncovered hundreds of cases in which public utilities have lied, cheated, and stolen and have been empowered to do so by the monopolistic status and government regulation they have enjoyed for so long. Our work is not easy; it is often drudgery aimed at proving minutiae or discounting a utility claim that is contrary to the physical evidence. Still, it is necessary work, and it provides those who are willing to learn with valuable insight into how electric utilities actually operate.

Every case story used in this book is absolutely true, unbelievable though some of them may seem. In almost every case, and with almost every client, we were required to provide the utility with extensive documentation in order to prove our point. Most often, the documentation was generated in some part by the utility itself, and you would be amazed at how many times utilities tried to deny the validity of their own documentation!

Over the years, we have been met with reactions ranging from outright hostility to amused disdain. More often than not, we are ignored, until it becomes obvious that we persist until we succeed. While we do not expect to be welcomed with open arms, we are always amazed when a utility, caught red-handed, defends wrongdoing and insists that we are the interloper—that everything would have been fine had we not shown up.

From the utility's perspective, of course, that is quite correct. Since the utility is almost always at fault, the situation we have brought to its attention might have gone uncorrected for all eternity, for all the executives cared. From their vantage point, things would have gone along swimmingly had we just crawled into a hole and vanished.

An attorney friend of ours once told us that we should never underestimate the level of hatred for our company on the part of the utility executives we had confronted. "If they hate us," we decided, "there has to be a reason." Here in Texas we have an old saying: "If it walks like a duck, quacks like a duck, and acts like a duck, then it is a duck." The same holds true for electric utility monopolies. The proof of their monopolistic status lies in the way they act and react and in their remonstrations to the contrary.

Nowhere to Turn, Baby!

Typically, regulated electric utilities have a specifically defined geographic area in which to transact business. Under most existing regulatory law, the power company is forbidden to service customers outside that area. But within the geographic limits, the utility has a monopoly. Under the scenario so commonplace in twentieth-century America, the utility has owned the market—100 percent of it.

Assuming that a utility does not do business exclusively in a rapidly expanding urban market, how do electric companies grow and improve financial returns within a finite marketplace? After all, the demands from stockholders and other vested interests are unremitting. In any business, investors must be continually satisfied that the business is doing everything possible to maximize profit. This would seem difficult for utilities, since they operate within a finite area and are precluded, by law, from entering into other business enterprises.

Of course, utilities operate with one important caveat: Unless their market size actually shrinks, the existing customer base remains static. Customers in a regulated area cannot simply leave or switch to another electric provider. That is, after all, the purpose of regulation. The law ensures that no one will compete with the power company's exclusive franchise.

But the utilities have developed some ingenious ways to maximize shareholder return while operating within regulatory guidelines. As we have seen, simply increasing the price of electricity is not the answer, because regulators have limited the utility's profit margin to a specific percentage of investment.

The next obvious thing to do, then, is to increase investment so that the specific percentage of profit will increase as well. Utilities buy airplanes and executive jets, and they build nuclear power plants, which engender huge cost overruns.

Naturally, such investment requires the approval of the regulatory agency. This is the point at which "schmoozing" the regulator plays such an important part in the utilities' success. We believe the overriding secret of regulated utility success lies in the ability of the utilities to effectively "get into bed" with regulators and influential politicians. Time and time again we see the legitimate needs and concerns of ordinary ratepayers shoved aside as regulators—those individuals charged with protecting the consumer—act as advocates for unwarranted utility profits.

Regulators Overwhelmed

In an era of shrinking state and municipal budgets, regulatory agencies are seldom adequately funded. States, in particular, simply have more pressing priorities—transportation and education initiatives always seem to have the upper hand. Regulatory groups get lost in the massive bureaucratic shuffle, and so does their funding. Of course, the power companies count on exactly that—and love every minute of the drama.

Only a decade ago, the total annual budget for the Texas Public Utility Commission was $7 million. And while that sounds like a lot of money to most of us, it's precious little with which to fund an entire agency, especially one charged with defending consumers and drawing the line on utility profits.

We watched with interest as the Texas PUC was quickly overwhelmed. Within a single year, three of the state's largest electric utilities filed separate motions for rate increase hearings—all at the same time. Indeed, "overwhelmed" is the word that best describes the situation. Consider this: To justify their rate increase requests, those three utilities spent more than $32 million in legal fees to outside consultants and lawyers—almost *five times* the entire budget of the Texas Public Utilities Commission!

The $32 million total did not include the time, effort, and energy expended by the utilities themselves, each of which had literally hundreds of employees pressuring the PUC to grant rate increase requests. When one public watchdog group accused the utilities of strong-arming the PUC, the utility spokesman responded that power company employees "were just trying to be helpful."

Around the country, public utility commissions and similar regulatory agencies have operated with surprising aplomb, given the tasks that they confront. We like to think of regulators as little boys with their fingers plugging the holes in the dam. Eventually, the sheer weight and pressure on the other side collapses the wall around them.

Being an overwhelmed regulator cannot be a pleasant experience, and we have enormous sympathy for the plight of those honest, decent, hardworking individuals who actively pursue the consumers' best interests. We believe, however, that the number of unbiased and unselfish regulators has dwindled at a rapid rate. The inevitable result is a system that has been rapidly co-opted and plowed under by the sheer weight of the utility juggernaut.

The real problem is this: Until competition among electric utilities takes a broader hold, regulatory agencies are the only advocate that consumers possess. If those boards and commissions knuckle under to utility pressure—which they are bound to do, given the sheer weight of force that most utilities will unhesitatingly bring to bear—consumers are left with no line of defense against the growing greed of the electric monopolies.

They Couldn't Care Less!

"Schmoozing" regulators to achieve specific utility goals often involves the application of large amounts of external pressure on a regulatory body. As we have seen, the regulatory body can only respond to the onslaught with rapidly dwindling resources. The real secrets of utility success, however, involve internal manipulation in three distinct areas: differentiation of the quality of service rendered based on the size of the customer, management of costs and inclusion of not-so-legitimate expenses, and the way in which some power companies succeed in "cooking" their books. These are the actual keys to power company growth and profitability.

Let's discuss the quality-of-service issue first. This is one front on which deregulation was supposed to make a substantial difference; after all, competing electric companies should, in theory, go out of their way to render excellent customer service. But from that perspective, deregulation typically makes things worse for the ordinary power consumer. Any utility provider, given the opportunity to compete for large accounts rather than small ones, will inevitably choose the former. That's just good business sense. The problem is that as the power generators try to develop a sort of critical mass very quickly, they offer terrific deals to larger customers and leave the smaller ones behind.

Existing utilities have a vested interest in defending their own "turf," and they spend inordinate amounts of time, effort, and money to retain large customers. Small users, which we define as anyone who pays less than $10,000 per month for electricity, get very little attention. In a best-case scenario, nothing improves for this group of ratepayers. The utility simply doesn't have time to bother with them.

Utilities manifest this lack of respect for small consumers in various ways. One popular tactic has been to send smaller power users a "simplified" monthly bill. Typically this bill contains only one thing: the amount owed to the utility. Included in the bill is a block of text that refers to a mysterious "Rate Code II," or some similar designation.

In the fine print the utility reminds the customer that detailed rate information is available at the utility's office. If you trundle yourself down to the local office, you'll be handed a notebook containing data for a score of different rates. But a search for "Rate Code II" turns out to be fruitless. It's probably just an oversight on the part of the utility, but "Rate Code II" is nowhere to be found.

Anxious to return home to your 2.3 children and the spouse waiting quietly by the light of the living room incandescent, you might finally ask a utility employee for help. "Where's Rate Code II?" The employee has to check with someone else, and so it goes. Finally, a half hour or so later, you learn the truth: "Rate Code II" is the same as "Rate Code G," which appears on page 47 of the notebook handed to you when you walked in and requested rate information.

The clear inference is this: Utilities frequently have their own rate codes, which are required in order to equate a billed rate code with the power company's own published rates. We often ask whether this strikes anyone else as a trifle odd. The only possible rationale for such rate code wizardry is that utilities genuinely do not want consumers to know how much "ordinary" customers pay for electricity. After all, ignorant consumers facilitate utility profits.

In point of fact, utilities would rather not bother with small customers at all. Anyone who sends in a monthly check for less than $10,000 isn't particularly important in the utility's scheme of things. Unfortunately for the utility, regulation (and, more appropriately, the demands of the competitive marketplace) requires that electric companies cater to all consumers, large and small. Clearly, the utilities don't exhaust themselves trying to hide their outrage at being forced to do so.

As we have said, going after larger consumers to the exclusion of smaller users makes good business sense for the power companies. And we would have no problem with the practice if performance matched utility rhetoric. But in addition to a widening discrepancy in service rendered (the biggest wheel, after all, gets the most grease), there's a wide credibility gap as well. Utilities actively seek ways to make certain that smaller customers are as uninformed as possible, while making use of consumer-oriented mass media to tout themselves as good neighbors and the saviors of every institution from free enterprise to Little League baseball. The reality is that the power companies couldn't care less.

"Revenue Enhancement"

Even though the utilities focus on large customers, the big power users aren't always well treated either. For more than a decade, our company chronicled massively creative "revenue enhancement" on the part of the giant corporation Entergy. One of Entergy's largest clients is the City of New Orleans.

As we mentioned, our company had audited the bills of a couple of large New Orleans–based consumers, the Superdome and the Sewer and Water Board. In both cases, we obtained substantial refunds from Entergy. Emboldened, we solicited and won a contract from the City of New Orleans to audit municipal electric bills.

Once we began to examine delivery of electricity to some of the city's buildings, we were puzzled. Some buildings had two meters—and received two bills—when one would have been sufficient. Of course, the two meters worked to the benefit of the power company, since some of the buildings failed to qualify for lower rates because their monthly usage was effectively split in two. Each building involved was being billed at the highest part of the declining block rate twice per month rather than once.

This practice violated the tariff. I had my first face-to-face contact with Entergy in the person of a PR professional—a person paid to "schmooze" politicians. "Oh, yes," he responded when we told him about the use of multiple meters. "That was just one of our revenue enhancement programs." The fact that such creative "enhancement" was illegal seemed to matter not at all.

Entergy's Big Easy

You'll be interested to know that Entergy refunded almost $500,000 to the City of New Orleans and about $200,000 to other clients of ours who had been victims of the same "revenue enhancement." But the next item on our agenda—examining the billing for streetlights within the city itself—would make "revenue enhancement" seem quite innocent in comparison.

Typically, street lighting is not metered; instead, the utility bills the municipality a fixed amount per light, per month. The billed amount can vary depending on the type of streetlight, wattage, ownership, maintenance, and so on. For the City of New Orleans, the total monthly bill was around $200,000 for some 50,000 streetlights.

We asked Entergy for an inventory that would identify each streetlight by location. Our intention was to compare the utility's list with the actual lights in service. Our request for the listing met with a good deal of righteous indignation—and we never got the inventory.

Undaunted, our staff drove around the city in a convertible with the top down, locating and documenting streetlight function and checking the lights against the bill Entergy presented to the city each month. We found that some lights for which the city had been paying a monthly bill were actually nonexistent. In other words, lights for which the city had been paying a monthly bill weren't even there. For other lights, Entergy had billed for wattage in excess of the bulb's capacity. Some lights had been double-billed. Additionally, some city-funded maintenance programs had been discontinued.

The greatest damage had been done by neglected maintenance. Entergy was supposed to repair and maintain the lights but had not honored that commitment. Initially, Entergy had employed a group of thirteen technicians to maintain the city's streetlights. Unfortunately, the unit had been disbanded in 1988, and streetlight maintenance effectively stopped.

By the early 1990s, even casual visitors to New Orleans could tell a difference. When presidential candidate Bill Clinton stopped at a fund-raising event in the French Quarter in early February 1992, he asked a member of his entourage, "Don't they have any lights down here?"

By the time of Clinton's visit, a large percentage of the streetlights throughout the city were, in fact, non-functioning. The city could not get the lights repaired; Entergy was supposed to be doing that work, and it wasn't getting done. Additionally, the non-functioning lights consumed no electricity, but the city was still paying for that unused power on a monthly basis.

For two years, we made an effort on behalf of the city to reason with Entergy. We finally gave up, drafted a lawsuit, and threatened to file the suit at the end of a thirty-day period if the utility kept silent. At 8:00 on the morning of the thirtieth day, the phone rang. On the line was the president of Entergy New Orleans.

"Can we work something out?" he asked.

Later that afternoon, we met with Entergy officials. The meeting lasted well into the evening. Amazingly, an entire room full of executives and lawyers actually came to an agreement that night on how the entire affair might be resolved. We drafted the terms on a single sheet of paper. In essence, after a new streetlight inventory was completed to everyone's satisfaction, the accounting firm of Ernst & Young would determine the amount of money Entergy owed the city.

Entergy's initial attitude was one of optimistic defiance. Some utility officials actually believed that despite the lack of maintenance and continuous billing without justification, the city might owe money to the utility. As the process unfolded, however, the executives learned that the outcome would be the other way around. Indeed, it began to appear that Entergy would owe the city much more money than we had imagined!

As the accounting emerged, it appeared that Entergy would owe the City of New Orleans about $12 million and change. The utility began looking for ways to shortcut the process and thus reduce the bill. The initial refund amounts that Entergy offered were only about a fourth of what was owed, but the dangling settlement appealed to cash-starved city administrators. The new mayor of New Orleans wanted to give police officers a raise, and he and his team had inherited myriad fiscal problems for which Entergy's cash would be a Band-Aid of sorts.

Entergy's offer to settle the case was about $3.5 million. The new mayor responded that he needed at least $6 million. After a good deal of rather bare-knuckle negotiation, the utility finally agreed to that figure. Entergy also agreed to restore all the inoperative streetlights, a process that could easily cost another $4 million to $6 million.

We work on a contingency basis. We get paid only after we recover money for our client. Suffice it to say that the new mayor didn't particularly relish the thought of writing us a check for $3 million. We eventually got $2.7 million and left town with a song of thanksgiving on our lips, thinking all the while that we would never return. Everyone involved—Entergy, the city officials, and our company—felt we'd gotten the short end of a very big stick. And our battles with Entergy had only begun.

Maximizing Costs and Expenses

As it turned out, Entergy's neglect of city lighting maintenance was just the start. The contract with the city had required the utility to replace *all* streetlights every four years. That commitment was breached by the utility in July 1988, when the New Orleans office disbanded the street-lighting department. Entergy's home office said street-lighting maintenance expenses were too high.

In its accounting of the streetlight debacle, Ernst & Young estimated that Entergy had saved more than $8 million by not replacing lights from July 1988 to May 1994. All the while, of course, the utility had continued to bill the city for the service that wasn't being rendered.

With surprising regularity, Entergy officials would appear before their regulators—the New Orleans City Council—and request rate increases to cover increased costs of doing business. Those requests were routinely granted. All the while, the utility was taking in vast sums of money from the City of New Orleans itself, much of it unearned.

We believe the only reason Entergy finally came to the negotiating table was because the truth was about to be revealed to the public at large. If that happened, Entergy would be pilloried in the court of public opinion. The resulting tidal wave of criticism—and the prospect of denial of rate increase requests—would hurt the utility far more than the $6 million settlement ever could.

Ironically, Entergy was able to keep its contract with the city. The arrangement continued just as it always had, although it was monitored a bit more closely. But after the settlement—or perhaps despite it—Entergy and the City of New Orleans returned to business as usual. If you had been doing business with someone who had deliberately cheated you for years and years, do you think you would ever entertain the notion of going back to the way things were? Would you?

Well, that's exactly what happened. The truth did win out—but for the truth to win, the truth had to be buried. Entergy suffered some pain, albeit for large gain. The city, satisfied that the utility could now be trusted to do the right thing, turned its attention to other problems. For us, the real mystery was why everyone in New Orleans seemed content to forget the entire affair.

The California Conundrum

What Entergy did to the City of New Orleans is endemic in standard utility practice. Costs and expenses, whether real or not, are inflated to support the utility's case for increased rates or state-subsidized bailout. The practice extends to the deregulated sectors of the industry, as California residents have seen all too clearly.

Truth to tell, some electric utilities thought they'd found a nirvana of sorts in deregulation. The State of California, for example, obligated itself to pay the utilities there billions of dollars for "stranded assets," or investments that would no longer pay for themselves in a competitive, deregulated environment.

Two years after signing what would have been a financial coup in any other industry, California's two largest electricity providers were filing for bankruptcy—and again California residents footed the bill. Now, power companies nationwide argue that they seek a newer, more equitable "model" for deregulation to follow. All they are trying to do is to slow the process down as much as possible.

California's deregulation disaster can be traced back to 1996, when the state's three biggest utilities banded together to force on ratepayers what Ralph Nader called "the largest corporate rip-off in American business history." Pacific Gas and Electric (which was then the nation's largest privately owned utility), San Diego Gas and Electric, and Southern California Edison were caught in a squeeze between their big industrial customers, which were threatening to generate power on their own, and the burden of their own bad investments in obsolete generators, mainly nuclear power plants.

The utilities were also tired of having their rates regulated by the state's ninety-year-old public utility commission. What the utilities wanted was nothing out of the ordinary; they just sought to cash out of their bad investments, keep the large customers, and generate massive profits without undue concern for regulatory agencies. And who wouldn't?

The three utilities proposed that regulation of distribution lines would remain intact. In other words, the utilities wanted to separate the business of generating power from the business of distributing it to the public. Clearly, the power companies intended to spin off much of their generating capacity. In point of fact, assets didn't exactly change hands; much of the transfer was only on paper. Power plants were transferred to the parent corporations of the distribution companies.

Once the utilities had become distribution companies, they could compete with other resellers for customers, who were then free to choose their suppliers. Amid burgeoning environmental concern, the utilities promised that consumers could even purchase "green" energy from companies selling wind and solar.

Competition will rule, the utilities promised. As a result of competition, advocates claimed, prices would go down.

But in exchange for lower prices, Californians had to pony up between $20 billion and $28.5 billion for "stranded assets." Actually, the utilities will argue that any cost is "stranded," in that a cost involves spending money rather than taking money in. These "stranded costs" were simply bad utility investments in unneeded or unproved generating facilities. The refund charges would be levied through "deregulation transition fees," which were effectively buried in customers' bills. These hidden charges added as much as 30 percent to the average consumer's monthly utility bill.

During the time required to reimburse these bad investments, retail prices for electricity were frozen. In the topsy-turvy world that is utility management, creating a static playing field while consumers paid for utility mistakes seemed the right thing to do. The California Public Utility Commission would also get $89 million in ratepayer money to promote the new scheme. The PUC's advertising blitz gave the existing utilities a considerable advantage over any other competition that might materialize.

In their haste to cash out, the utilities involved made critical mistakes. Most important was the basic utility assumption that an enormous supply of cheap wholesale electricity would always be available. Consequently, the utilities sold off too much generating capacity and retained too little of their own ability to supply demand for power. Then came a hot summer and a rare cold winter. California natural gas prices shot up, and some key generators went down. Storms knocked out transmission lines. Nuclear generating plants experienced a wide range of problems. Suffice it to say that the utilities found themselves at the mercy of independent producers—the generators who had snapped up generating capacity and could thereby manipulate the wholesale market.

The utilities realized too late that consumers, who were buying power at fixed costs, had little incentive to conserve electricity. Consequently, demand quickly outstripped the supply of cheap wholesale electricity. In a seller's market, wholesale prices rose at the whim of those with power to spare.

Companies like Duke Energy, Reliant of Texas, and Enron made billions of dollars in profit selling power at high rates to the same companies that had just given up their generators. The rate freeze still in effect created some odd dichotomies; for example, the three companies were forced to sell power to consumers at a rate of $64 per megawatt-hour while paying $1,400 for the same amount.

Even rival utilities got into the act. Oregon's Portland General Electric withdrew a proposed rate hike for its own customers when it realized it could sell power in California at a higher profit. At least two large bauxite smelters in the Northwest shut down and realized some $500 million in profits by selling cheap electricity they had already bought via long-term contracts with hydro generators. Selling power was, simply, more profitable than making aluminum.

Interestingly, the parent companies of the California utilities that had begun the entire fracas made as much as $3 billion selling power to electricity distributors, many of whom were now pleading for state help to stave off bankruptcy.

California governor Gray Davis made dozens of calls to the Federal Energy Regulatory Commission and other federal agencies in an effort to fix prices, guarantee the supply of electricity to California, and punish the companies like Enron, which were price-gouging California consumers. Of course, Davis hadn't reckoned on the massive political muscle inherent in the mammoth energy companies. Federal agencies were simply unable to rein in the powerful suppliers.

California's state and private utilities were caught between the proverbial rock and a hard place. San Diego Gas and Electric, which had fewer stranded costs to pay off and thereby quickly escaped the rate freeze, doubled and tripled rates. Understandably, the rate hike infuriated Southern California consumers. California's state government rode the back of the taxpayers, buying power to resell to SoCalEd and PG&E in order to save the massive utilities from bankruptcy.

Remember that all this was necessary purely because utility rates were frozen. But if rates hadn't been frozen, they would have doubled and tripled, thus infuriating the rest of the state.

We'll discuss the outcome of the California fiasco in the next chapter. At this writing, none of the other states that have deregulated have suffered a disaster on the scale of the California debacle, but deregulation's track record thus far has been largely unimpressive for all concerned.

Accounting Alchemy—It Didn't Start with Enron!

In essence, the California distributors created a massive statewide panic by threatening bankruptcy, a threat substantiated only by losses on paper. Meanwhile, their parent companies were quietly generating something of their own—huge profits.

We believe most of the larger companies never bothered to adequately account for all the stranded-cost money. All that cash is probably tucked away in foreign and out-of-state investments. Meanwhile, the rate-paying public continues to get the shaft.

During the Middle Ages, so-called scientists schemed to render gold from more ordinary substances, failing to understand that if they succeeded, the gold they produced would be so plentiful as to be worthless. These early experimenters were called "alchemists," and their craft was "alchemy." The utilities practice a modern-day version of alchemy, with one important distinction: Unlike the alchemists of old, the utilities make alchemy work. They create gold from paper and ink.

"Accounting alchemy" is not new to the electric utility industry. We found evidence of the same sort of shell game in New Orleans several years ago, when we discovered that Entergy was carrying as an asset the same streetlights the company had sold to the city some years earlier. You'll recall our discussion of "The Second Battle of New Orleans" in the introduction to this book.

Fudging the books is not a new art. Remember Samuel Insull, Thomas Edison's business partner, who convinced the City of Chicago to create the first electric monopoly? As the first real utility executive, Insull garnered another singular distinction: He was the first utility crook. Unfortunately, we don't remember the story today; it unfolded during the Great Depression. Still, Insull's tale bears startling similarity to recent debacles in California and southeast Texas. So indulge me in a quick history lesson. The point we are making will become evident as we move along.

Insull was born in England and moved to the United States in 1881, where he became Edison's assistant. Insull was such a financial wizard, as the story goes, that he'd memorized Edison's European financial affairs ledgers and was thus able to save Edison something like $150,000 his first day on the job. By the time Insull came on board, Edison had already begun to put his utility empire together; he had already constructed the first electric power plant in New York City.

Edison was a genius, but he was not generally known for his organizational skills. In hindsight, it's easy to understand how Edison might have been quickly convinced to make Samuel Insull his business manager.

Edison Electric eventually became General Electric, courtesy of J. P. Morgan, and Insull was finally forced out by the new owners. Undaunted, he headed west to Chicago, where he quickly built an empire in the burgeoning power business. Indeed, Samuel Insull was something of an inventor in his own right, having designed the world's first steam-engine turbine. As one biographer noted dryly, "Working with Edison, Insull had obviously picked up a trick or two."

From his Chicago base Insull developed the distribution mechanism that would bring electric power to rural areas. In 1912, the year the *Titanic* sank, Insull created Middle West Utilities, which would eventually control one-eighth of the entire American electric power marketplace.

As America entered the postwar boom that would become the Roaring Twenties, the electric power industry was dominated by three enormous utilities: the Southern Company, which spawned, among others, utility giants like Duke Power; Insull's Middle West, which also operated in southern Canada; and the Morgan Group, which included companies such as Consolidated Edison and Public Service of New Jersey.

The era of the Roaring Twenties was essentially a celebration of the power of American private enterprise, particularly business and industry. Some observers, however, were able to discern the truth behind all the hoopla. Senator George Norris of Nebraska said that the electric power companies represented the "greatest monopolistic corporation that has been organized for private greed." The senator went on to complain that the utility monopolies had "bought and sold legislatures . . . and interested itself in the election of public officials, from school directors to the President of the United States."

Based on our quarter century of experience auditing the electric industry, we'd say nothing has changed.

Insull made enormous contributions to the Republican Party, all the time pretending he had no thought of buying political influence. Indeed, the campaign contribution situation became such a monumental issue that the chairman of the Illinois Commerce Commission, Frank Smith, was prevented from taking a United States Senate seat he had already won.

For Insull to expand his utility empire, it became necessary to borrow huge sums of money. This was a problem because while serving as Edison's partner, Insull had grown highly suspicious of the whole Wall Street investment banking crowd. "Bankers will lend you their umbrellas only when it doesn't look like rain," he once observed.

Consequently, Insull chose to deal locally with banks like Continental Illinois. Insull's biographer, Forrest McDonald, notes that building a pyramid (as Insull would do with his holdings) was as easy as 1-2-3. McDonald recalls a Chicago banker who once saw Insull at a party and said, "If you fellows ever want to borrow more than the legal limit, all you have to do is organize a new corporation. We'll be happy to lend you another $21,000,000."

Indeed, Insull's bookkeeper remembered receiving calls from bankers "the way the grocer used to call my mamma." The bankers, the bookkeeper noted, would "try to push their money at us. 'We have some fresh green money today, Mr. Insull. Isn't there something you could do with $10,000,000?'"

Insull's power business required huge outlays for plants and equipment. But the New York financiers were losing out on the business to the locals in Chicago; back in the 1920s, as now, financiers didn't take that sort of thing lightly. As one of Insull's bankers said, "These New York fellows were jealous of their prerogatives. If you wanted to get along you had to be deferential to them and keep your opinions to yourself." Apparently, old man Insull wouldn't do that, thus manifesting one of the few reasons to think highly of him.

Insull operated his power-generating businesses skillfully. His plants were among the most efficient in the nation, producing electricity at rates that were less than half those of his competition. Insull, unlike some of his modern-day counterparts, was reputedly a generous man. He treated his employees with kindness, buying back shares of stock from time to time to distribute among the workers.

But the problems with New York persisted. Wall Street's representative in the area, Cyrus Eaton of Cleveland, began buying up shares in all of Insull's operations. Fearing that he was about to be taken over in a hostile bid, Insull arrived at a rather novel way to protect his holdings. He kept consolidating his companies and entangling them, one inside the other, each loaded with debt. The aim of the elaborate shell game was to make Insull's businesses less attractive to outside investors. In December 1928, Insull and his Chicago-based investment bankers formed a new company, Insull Utility Investments, which owned large blocks of shares in all the companies under the Middle West banner. To effectively block the strategies of the era's corporate tycoons, Insull acquired a majority of the new firm's stock.

The stock, incidentally, went on the board at $12 a share and finished the first day at $30. By the spring of 1929, it was $150 a share. Insull's other holdings soared as well; Middle West Utilities shares rose from $169 to $529 during the first eight months of the year.

Insull's financial success meant that his shell games continued unabated. He and his investment bankers organized another company, Corporation Securities of Chicago. Eventually, the entire Insull complex had booked assets of more than $2.5 billion, and his electric empire served more than 4.5 million customers.

The Wall Street crash, in November 1929, brought to a close the first era of "cooked books" high finance. The second such era is still ongoing. Nearly every high-flying enterprise took a tumble in the Great Depression, and Samuel Insull's business fell further than most.

While Insull's stock held up well in the weeks immediately following the stock market crash, it had certainly been weakened by the same economic catastrophe that had brought the entire world to its knees. New York financiers sought revenge in the form of a "bear raid," betting that the stock price would fall. Eventually, Insull's spectacular pyramid collapsed, as the Depression caused a dramatic falloff in the power business, and Insull couldn't service his huge debt load. One banker called him "too broke to be bankrupt."

Insull was eventually charged with mail fraud and embezzlement, among other things. Just as Enron chairman Kenneth Lay fled to his Houston apartment for sanctuary, so Insull fled to Greece, which didn't have an extradition treaty with the United States. Eventually, political pressure forced him to return to the States, where, ironically, he was exonerated of all charges. What saved Insull? His attorney had him relate his life story to an enraptured jury. In the end, the charges didn't matter; Insull's personality saved the day.

Of course, Insull's reputation was forever ruined because he fled to avoid prosecution. Out of hundreds of thousands of stockholders—all of whom lost most or all of their investments in Insull's companies—one would think enough could be found to form a lynch mob of sorts. But that never happened, although one might argue that the old man got what he deserved. Samuel Insull died penniless in a Paris metro station. He was seventy-nine years old.

Today, with accounting laws and regulations far stronger than they were in the 1920s and 1930s, Insull would have needed more than charisma to save himself.

"Today," one observer noted, "Insull would have done hard time." Ironically, Insull's quality of life might have been better had he gone to the slammer.

One positive result of Insull's debacle was the Public Utility Holding Company Act of 1935, which required all holding companies that owned public utilities to register with the Securities and Exchange Commission. The act gave the SEC the authority to break up large utility empires. But the big hammer has always been carefully hidden; outgoing chairman Joe Kennedy said at the time that the commission shouldn't be allowed to act this way. Many would argue that it never has.

3

THE CHALLENGES OF DEREGULATION

Senator Johnson's Little Bill

Ohio is a bellwether state. The people there are progressive, in a uniquely conservative sort of way. You may have heard the old saying: "As Ohio goes, so goes the nation."

Let us hope that adage is more homily than truism.

While Ohio lawmakers disagreed about the potential savings inherent in deregulating electric utilities, they did act to bring about a fundamental change in utility operation. After giving Ohioans multiple ways to bargain for cheaper rates, Ohio lawmakers voted to end the monopoly for electricity suppliers.

In late June 1999, the Ohio Senate voted 29–3 to approve a revised plan to deregulate the state's electric utility industry. Bob Taft, the governor of Ohio, was a strong supporter of deregulation and signed Senate Bill 3 into law.

The author of the bill, Senator Bruce Johnson, said, "This bill may be the best consumer package in the country." His contention that residential consumers and small businesses would enjoy the savings expected by large industrial users had been challenged by opponents of the bill.

Skeptics like us maintained that critical decisions affecting the chances for real savings would be made by the Public Utilities Commission of Ohio, an agency that had already approved the high rates for electricity being paid by many northern Ohio communities.

One of Johnson's counterparts, Senator Leigh Herington from Rootstown, was rather blunt in his assessment. Senator Herington urged caution: "Will everybody be able to benefit? We hope so. Can we guarantee it? Not as we sit here today."

Herington voted for the bill, but encouraged lawmakers to keep a close watch on PUCO to ensure that the panel made fair decisions on how the current investor-owned monopoly electric suppliers were compensated for opening their markets to competition.

Other Ohio senators, like Greg DiDonato, a Democrat from Dennison, expressed little confidence in PUCO and voted against Johnson's bill. He said southern Ohio, which had enjoyed relatively cheap electricity costs, might be damaged by commission decisions aimed at opening the market.

Beginning in January 2001, Ohio abolished the service territories for the eight monopoly suppliers—territories carved out nearly a century before. The new law allowed customers to choose their electricity provider. Only the generation of electricity was being deregulated; the current monopoly continued to transmit the power to consumers and maintain the system.

The interesting feature of Johnson's bill was a provision that the senator added in order to sway other lawmakers. This clause forced suppliers to reduce the cost of electricity by 5 percent of the current cost to generate power, which amounted to $1 to $2 per month for the average Ohio residential customer. One observer called Johnson's 5 percent guarantee "only the tip of the long-term iceberg."

In Ohio, steelmakers represent the state's largest group of electricity consumers. The steel companies expected to save money on future power bills once Johnson's bill was enacted into law.

As you already know, however, large buyers typically get volume discounts and thus would not see lower rates immediately. On the other hand, smaller companies that supply services and materials to steelmakers should see a decrease in electricity costs, and might choose to pass those savings on to larger customers, like the steel companies. Consequently, auto, appliance, and home-building customers might get steel cheaper from the company and pass those savings on to retail consumers.

Johnson's bill also gave cities and counties the right to purchase electricity on behalf of residents in their communities, unless local residents decided to "opt out" and choose their own suppliers. By purchasing on behalf of a large number of small consumers, city and county governments expected to get power at a cheaper price.

Additionally, the bill allowed small businesses and related organizations, such as schools or hospitals, regardless of location, to bind themselves together in a sort of geographically widespread "buying pool." The volume-discount logic held that these associations would also be able to bargain for cheaper electric rates.

Consumer groups candidly doubted that Ohio-based businesses would pass their savings on to end-user consumers, but they hailed the passage of the bill because it gave Ohio residents the opportunity to band together and bargain for lower prices. This unique feature of Johnson's legislation had been lacking in the deregulation of cable television and telephone service, and those industries had not reduced their prices significantly.

At the time, most consumer advocates believed that benefits to consumers would come from local government officials working to build support for organizing residential buying pools. Through a multitude of these arrangements, "insignificant" residential customers could draw the attention of electricity providers who were willing to offer cheaper rates to groups or "blocks" of individual consumers.

At this writing, the Ohio experiment in deregulation has been ongoing for just over five years. As with anything else new and unproved, the results have been mixed. By November 2002, for example, Columbus residents were becoming disenchanted with deregulation, largely because of the exorbitant "connection fees" incurred by American Electric Power Company customers.

New homeowners were being charged a one-time $375 "connection fee" and an $8 monthly surcharge to obtain service, while Ohioans moving into new apartments paid a $4 monthly surcharge. These fees were—you guessed it!—part of an agreement with the utilities approved by the Public Utilities Commission of Ohio.

The surcharges, in most cases, effectively wiped out the 5 percent savings on electric bills guaranteed under Johnson's bill, which had taken effect only a year earlier. The utilities needed less than twelve months to completely undo any vestige of potential for consumer savings.

During early 2002, AEP and other utilities began charging about $1,500 per new hookup, ostensibly to cover the expense of extending service from the pole to individual homes and apartment buildings. Some consumers complained to the PUCO about being charged connection fees ranging as high as $15,000! To its credit, the commission did order that those excess charges be refunded, but it did so only under pressure from watchdog consumer groups.

The executive director of Ohio's Building Industry Association said that although connection charges are usually paid by builders, property owners will pay them back—one way or another. One realtor explained the surcharge this way: "This has to do with the cost of building a new house. It has nothing to do with the cost of energy." In actuality, the charges have more to do with helping the utilities circumvent a competitive marketplace than anything else. In the Buckeye State, electric utilities operate true to form.

The connection fee surcharges are scheduled to continue until the end of the Ohio utility deregulation transition period, which means that some customers will continue paying the exorbitant fees until the end of 2008. We imagine those folks are waiting for 2009 with great anticipation!

Before deregulation, of course, utilities routinely picked up the cost of extending electricity service to new dwellings. The so-called connection fee was considered a cost of doing business and was built into base electricity rates. No longer. When the government taketh away from the utilities, the utilities find another way to giveth back to themselves.

Consumer advocates complain about these "hidden costs" of deregulation. The sad fact is that if deregulation worked as it was designed to work, there would be no hidden costs. If governmental regulatory entities like the one in Ohio actually acted on behalf of the consumers they were appointed to serve, "hidden costs" would be nonexistent.

Someone once said that the definition of insanity was doing the same thing in the same way again and again and expecting to see a different result. If we apply that logic to Ohio's deregulation experience, we might conclude that deregulation is less than worthwhile. Exactly the opposite is true: Utilities never negotiate or give ground on profit-oriented issues; to expect them to do so is pointedly irrational. What is required is a system that caters not to the power companies but to the consumers being held hostage.

An Industry in Flux

Call us overly optimistic, but we believe that something—some new technological advancement or a fundamental paradigm shift on issues relating to government interference in business—can produce a system in which utility customers are given precedence over the interests of the utilities themselves. We just haven't found that something yet!

The utility industry as a whole may be in a state of flux, but the momentary confusion isn't the result of deregulation. While increased competition has become an attractive idea, the notion is popular not because it holds the power to rein in electric utility excesses but because our fundamental attitudes toward government and business are changing. Guilty though they might be of everything else, the utilities themselves have done nothing to foster this paradigm shift.

Indeed, deregulation has not yet proven to be a panacea. Instead, the introduction of competition into the utility marketplace has made necessary new governmental entities (or at least new entities overseen by local and state governments) aimed at ensuring that all participants play fairly. Like a teacher patroling an elementary school playground, these states have discovered through their own trials and tribulations that deregulation is most often not a hassle-free stopping point for monopolistic utility practices. California and other states have learned this lesson the hard way.

The "restructuring" that the utilities favor has no real parallel to the intent of deregulation. While deregulation will likely continue in terms of price scale and the entry of competition into the utility marketplace, government intervention at some level is likely to continue also.

Of course, the status of this "restructuring" process varies considerably from state to state. At the time of this writing, more than half of the states have passed legislation or enacted regulations designed to effectively restructure their own electric power industry. Statistics from the U.S. Department of Energy show that those states with higher-than-average electricity prices, such as Pennsylvania, New York, and the rest of New England, have successfully opened their retail electricity markets to competition. Only time will tell whether allowing customers to choose their own provider will generate an appreciable cost savings for the consumer.

Interestingly, state restructuring legislation or regulation has either encouraged or mandated that the utilities divest themselves of generation assets. By the end of 2003, more than a quarter of all utility generating capacity had been sold to unregulated companies or—and this is an important caveat—transferred to unregulated subsidiary companies. In some parts of the country, like much of New England, almost every generating plant has been sold to independent power producers. Today, these independent generators produce almost half of all wholesale electric power in the United States.

Deregulation—Or Restructuring?

Obviously, the electric utility industry as a whole is shrouded in uncertainty. Deregulation has certainly had an industry-wide impact. The utilities are gradually moving toward a structure in which competitive utilities will generate the electricity, while utilities continue to hold transmission and distribution facilities in a monopoly-like grip. But restructuring holds sway in many states; rather than endure what Californians suffered a few years ago, most state legislatures are leaning toward "restructuring" instead.

Witness the latest round of changes in the state of Oregon, where customers of both Portland General Electric and PacifiCorp were given their choice of provider under the state's electric industry restructuring law. In Oregon, as in so many other states, the wheels of deregulation have been in motion for some years; the state's largest investor-owned utilities were "restructured" back in 1999.

The 2003 restructuring law was designed to give utility customers more retail options while encouraging the development of a competitive energy market. As has become accepted practice, the utilities with transmission and distribution apparatus already in place will continue to deliver electricity, while consumers are now empowered to choose the company that will produce their power.

Oregon's restructuring law is actually a combination of new legislation and the administrative rules adopted by the Oregon Public Utility Commission. The major concern of the utilities was covered by the PUC rules, mandating that large electricity customers continue to purchase power from their current utility. The customers up for grabs are the smaller business customers and residential consumers, about whom most large utilities couldn't care less.

The PUC also mandated that a 3 percent "public purpose charge" be collected from all retail customers. The charge is tantamount to extortion; moneys collected are supposedly used to fund and encourage energy conservation and to pay for the costs of development of renewable energy. Both causes, of course, continue unabated at a snail's pace.

More than half the surcharge money—57 percent—has been dedicated to loosely defined "conservation education," while only 17 percent has been allocated for development of renewable energy. Nearly 12 percent is applied to low-income weatherization (which, it might be argued, is conservation education targeted at the portion of the rate-paying public who need it most), and about 5 percent is designated to benefit low-income housing. Specific application of these funds is left largely to another newly created entity, called the Oregon Energy Trust. Why the PUC decided that commonsense effort required a 3 percent surcharge is frankly beyond our understanding; doubtless the move was aimed at saving the power companies a little money.

While Oregon's law did establish a general framework for the future of utilities in the state, much of the implementation of the new statute was left up to the state's PUC through the usual rulemaking and rate-setting processes. And the PUC, as might be expected, has largely opted to act in behalf of the established utilities. For example, utilities in Oregon aren't required to sell any assets that generate electricity, and the PUC has mandated that no consumer should be "forced" to choose from competitive providers. In other words, competitive utilities entering the marketplace are required to "sell" themselves to consumers. Otherwise, the entire consumer base would be something akin to a large herd of free-ranging cattle. The law's application makes it a challenge or chore for customers to change providers, and the utilities are smart enough to realize that a majority of retail customers, when faced with the challenge of changing providers, will make the choice that requires the least possible effort on their part.

What Price for Efficiency?

The utilities evidence a studied indifference to ways by which their product might be produced more efficiently. This fact flies in the face of the very real threat faced by power companies: the threat of losing their largest customers. Larger industrial customers represent a substantial income base for the utilities, and the loss of more than a couple of large consumers could have a serious effect on a utility's financial statements.

For years, customer-side efficiency programs, called demand-side management programs or DSMs, have been trotted out by the utilities as evidence of a good-faith effort on the part of the power companies to help large customers reduce their electricity use and demand. At best, these programs have helped customers cope with steady increases in utility rates; at worst, they have been ineffective. Today, however, DSMs are undergoing rapid and substantial changes in almost every marketplace in which they are offered.

DSMs had their start at the behest of overwhelmed regulators, who saw them as an easy solution to utility requests for construction of new power generation, transmission, and distribution facilities. After all, the reasoning went, utilities ought to take an active role in reducing demand for their services. Unfortunately, things haven't exactly worked out that way.

Why not? First, because regulators, in cooperation with the utilities, decided that the cost of DSMs would be borne primarily by—you guessed it—ratepayers. Imagine that! Second, there is the issue of resource allocation. As utility markets become progressively more competitive, price, rather than efficiency, becomes the central issue. Many utilities (and indeed, some larger consumers) see the DSMs as an unnecessary use of time and resources.

Third, only a few utilities have been clear-thinking enough to see DSMs for what they are—opportunities to retain larger customers through increased efficiency and reduced costs. Those utilities that have made effective use of DSMs have seen customer retention increase, and the programs themselves have fostered economic development and new business opportunities. Today, however, the cost for the effort is usually borne by the utility itself, which explains why the initiatives are seldom implemented and even more seldom work.

The motivations for industry-centered DSMs *are* slowly changing, particularly as deregulation makes it imperative for the power companies to offer something of greater value in order to retain larger customers. DSMs clearly offer an opportunity for substantial energy savings; in many situations, the savings gained from proper implementation of DSMs may exceed the savings inherent in utility restructuring. Thus have the utilities managed once again to shoot themselves in the foot. By ignoring a need for viable DSMs, they have actually increased the chances that their largest customers will, when given the chance, opt to find another provider.

You may wonder why the power companies have largely ignored the marketing opportunities inherent in DSMs. We believe the utilities' attitude toward DSMs centers on the very nature of such partnerships, which would force the utility to work directly with larger customers. Such one-on-one contact necessarily implies that the electric company sees a genuine value in continuing to do business with the customer. As we have seen, such attitudes among utilities are relatively rare.

The notion of sharing risk and reward for mutual benefit and survival is a fairly recent innovation for the power companies. In decades gone by, the utilities shunned almost every opportunity to increase their market viability by meeting specific customer needs, just as most electric companies have ignored the opportunity to secure long-term business with ten-year fixed-rate contracts and other commonsense devices.

The other opportunity that the utilities have largely ignored, for whatever reason, has been technological advance. For too long, the typical method of producing and marketing electricity has included large central power plants to generate electricity, with customers linked to the utility through a network of wires. Any other system of generation and delivery has been considered too outlandish or speculative to warrant much attention. But that ancient attitude may change quickly as new technologies serve to beat the utilities at their own game.

New technology offers the promise of decentralized power supply and disconnected users, albeit in a positive, the-lights-are-still-on sense. For example, fuel cells can generate both electricity and water, while micro-turbines powered by cheap natural gas are being coupled with photovoltaic cells and energy storage systems to produce electricity from the sun. Such devices are in limited use today, but as awareness grows, the advocates of new technologies are marketing them to an increasingly large portion of the utilities' customer base.

Consequently, new technology offers the promise of freedom for customers at every level—freedom from the tyranny represented for so long by an uncaring and disdainful electric monopoly. To date, the power companies have done a good job of casting alternative-power experimenters and suppliers as foolish modern-day alchemists unworthy of notice. What these individuals are touting, we are told, is eccentric, far-fetched, and unreliable. Unfortunately for the utilities, nothing could be further from the truth. The simple fact is that these alternate sources aren't being utilized to their fullest potential, and the utilities are doing all they can to prevent widespread adoption and acceptance.

President John F. Kennedy once observed that advances in technology will affect us whether we desire them to or not. The new power generation systems should serve as a wake-up call for the power companies, but instead the proponents of new methods and new technologies are roundly derided and airily dismissed. What goes around does eventually come around, however. Within a few years the utilities may well find themselves at the mercy of the same advancing technologies they now disdain.

California—The Real Deal

But new technology gets short shrift as consumers begin to deal with the tangled mess left behind by ill-conceived and poorly implemented deregulation schemes. As you'll recall, the California energy deregulation plan was designed as a simple initiative to open up the market, increase supplies, and lower rates for consumers. The impact of the plan, however, has been anything but simple. In fact, the plan became a political snowball aimed squarely at the incumbent governor, Gray Davis. The fallout is still affecting the legislators who approved it, the state and federal regulators who oversaw it, and the consumers and businesses who now must pay a steadily rising state energy bill.

Interestingly, many of the solutions originally embraced by energy deregulation in California seemed to have created more problems than they solved. For a time, Governor Davis pushed a plan to stave off utility bankruptcies by using billions of dollars of state money to buy up transmission lines, but the state legislature balked and political opponents seized on the idea that Davis was overpaying for an antiquated system. That was probably true, but Davis did help stabilize energy prices. All this was done at a cost, as the governor spent billions from the state treasury to buy power.

Additionally, the long-term power-purchasing contracts to which Davis committed now seem something of a bad deal. Consumer groups claim that the contracts burden ratepayers with locked-in high energy prices for decades to come. One positive note: The long-term contracts, a milder-than-expected summer, and conservation efforts by the state's citizens did combine to create a rather temporary power glut in the state. Unfortunately, the surplus energy bought under the long-term contracts had to be resold at one-fifth of the cost the state paid.

The financial loss in one month alone was estimated at $46 million. A long-term power surplus (for which everyone wished some years ago) cost the state billions of dollars in excess power. The only upside was that the surpluses seemed to have broken the runaway market that caused energy prices to spiral in the first place.

Deregulation's convoluted aftermath created a raft of other complications. Davis came down rather hard on out-of-state power suppliers, when he and others claimed that out-of-state companies had pillaged the state treasury for billions of dollars in excessive energy charges. The companies themselves brought the negative spin to a screeching halt by threatening to stop investing in the new power plants that California requires for future growth.

While the heated political disputes over California energy deregulation may seem parochial in scope, their eventual impact on the industry as a whole will be much more wide-ranging. Davis, casting about for another scapegoat in the midst of a fight for his political life, accused the Federal Energy Regulatory Commission of allowing "unjust and unreasonable rates to prevail for months without taking action."

Scapegoating it may have been, but Davis's comments succeeded in pushing the commission to act. The agency slapped electricity price limits on California and nine other Western states and convened an unsuccessful two-week-long summit to resolve issues between California and its power suppliers. One may argue that much of this was intended as a rather splashy show; all Davis had done was rouse the utilities' ire and help power providers develop a short-term plan to regain control of the political situation.

All this was lost on the governor, because what he really wanted was a refund from the power suppliers. More specifically, he sought a check for the $7.4 billion he claimed was overcharged through market manipulation. He also asserted that the federal government had not been doing enough to help California in its time of crisis. The administration countered that the state should be building more power plants and stop trying to shift the blame to the federal government.

One might argue that Governor Davis's expectations did little to help him in his battle to retain his job. Two things should have been obvious to him early on: First, he wouldn't get any money from the power companies, and second, his actions would arouse some righteous indignation in the nation's capitol.

Political squabbling aside, many difficult issues were left resolved. For example, California stepped up to the table with the aforementioned $7.4 billion in energy costs to keep the juice flowing to the utilities. Who would wind up paying this multibillion-dollar energy deregulation bill? The utilities and their parent companies, of course, disclaimed any responsibility, while California taxpayers, residential consumers, and larger utility customers believed it to be an unfair burden. In the end, the consumers lost.

Essentially, the $7.4 billion question sparked another series of debates: Should the state of California create its own power authority and generate its own electricity? Should California's electric utilities be allowed to go bankrupt? Had outside suppliers routinely overcharged for energy during California's crisis?

State regulators decided to address none of these issues. Instead, they divided the $7.4 billion in "above-market" charges among the customers of the state's three investor-owned utilities. Consumers were stuck with the costs of long-term electricity contracts the state signed during the 2000–2001 power crisis.

The PUC was faced with four complex plans developed by the state's Department of Water Resources. All the schemes sought to find "equitable" ways to divide billions of dollars of power purchases. On a 3–2 vote, the PUC shifted $733 million in costs to San Diego Gas and Electric, a subsidiary of San Diego–based Sempra Energy.

The *Los Angeles Times,* on one of its more clear-headed days, said the panel's action made for "an expensive day" for SDG&E customers.

Consumer groups are bound to be upset by the regulatory action, which took place just days before this writing. Many Californians believe that the power providers (and, by consequence, their financially healthy parent companies, which have been freed by deregulation to trade energy and invest in power production outside of the state) should foot the bill. The idea of throwing this burden onto the consumer was the main impetus that launched a statewide voter initiative to turn Governor Davis out of office. Now that he's gone, regulators have opted to act in behalf of the utilities and against consumers.

What's next? We expect a new initiative to re-regulate California's power industry. In our view, energy deregulation is unwinding in California and other states because unregulated energy markets are easily manipulated by the high-stakes power providers. As energy costs continue to rise and the massive long-term contract costs are passed on through higher utility bills, ratepayers wind up paying billions for the energy the state purchased on the utilities' behalf. It's hard to imagine that voters, who were promised lower bills when the state deregulated energy years ago, won't decide to go back to the way things were.

The unique aspect of this entire affair lies in the fact that for some years California has had the mechanism in place to operate power plants. While we do not expect that the state will ever build, acquire, purchase, or seize power production facilities from regulatory-shielded utilities, the issue will continue to be the subject of much debate. Such an action would have a dramatic "ripple effect" on the energy deregulation movement nationally.

Some years ago, California actually created a power authority that could run the plants. Consumer advocates and some state legislators believed (and many still believe) that such an approach could solve the state's energy woes.

The notion of a state-run utility has its detractors. For example, former secretary of state Bill Jones, who led the charge that eventually sent Governor Davis packing, argues that the state should stay out of the power business if at all possible. In a keynote address before the Western Power Trading Forum, Jones slammed Davis for allowing the energy issue to evolve from a manageable problem to a crisis because the governor "lacked the will to make the tough decisions that needed to be made."

"There were some technical flaws in how deregulation was set up," Jones said, understating the fact. "These flaws were magnified by a temporary but extreme shift between supply and demand. The population increase in the rest of the Western energy grid combined with drought in the Northwest shut down much of our normal supplies.

"You folks are going to have to make a few changes to the way you do business in California," Jones told his audience. "The political climate has changed dramatically and either you get with the program or Californians may take matters into their own hands."

If this was a velvet hammer of sorts, the power providers obviously didn't grasp the subtlety. As long as California's regulators continue to pull their fat from the fire and as long as there is no large-scale ratepayer revolt, the utilities don't have much to fear.

The entire California debacle, in hindsight, was the result of an Enron-sponsored attempt to break up monopoly control of the state's energy markets by the local utilities. Enron Corporation and Kenneth L. Lay, its former chairman, were about the most persistent lobbyists anyone had ever seen in California.

With lavish campaign contributions, a fleet of lobbyists, personal pitches by top Enron executives, and powerful politicians in their pockets, Enron was the biggest player in the state legislatures in the late 1990s, pushing its version of energy deregulation—and nowhere did Enron push harder than in California.

Of course, most of the focus on the Enron scandal has centered on the company's efforts to influence Congress and the White House, but Enron had been conducting a similar campaign on the state level, with far less visibility and far greater success.

Enron's objective was to break up monopoly control of energy markets by local utilities and change the rules so that energy would be deregulated. During the three-year period from 1997 to 2000, nearly half the states adopted some form of energy deregulation, allowing energy companies like Enron to flood into virgin markets.

"Enron was . . . everywhere," says Paul Joskow, director of the Massachusetts Institute of Technology Center for Energy and Environmental Policy Research. "They not only carried a lot of water for themselves. But they carried water for the rest of the utility industry as well."

So determined was Enron that it frequently dispatched top executives like Kenneth Lay and Jeffrey K. Skilling, its former chief executive, to meet with utility commissioners, testify before statehouse committees, and call on local politicians. Of course, all the gladhanding was backed by hefty campaign contributions. Enron gave nearly $2 million to more than 700 candidates in 28 states during the same three-year period.

Indeed, in the 2000 election cycle alone, Enron gave more than $1 million to local candidates. The contributions ranged from as little as $250 to $500 in hundreds of districts to as much as $10,000 to influential state legislators in California.

The phrase "bought and paid for" does come to mind.

Unlike utilities or consumer groups that lobbied for deregulation in one state or a few, Enron took on the issue nationally. Its unique strategy combined a promise of lower electrical costs with old-fashioned statehouse power politics. The latter worked; on the former issue, the jury is still out.

During the mid- to late 1990s, contributions streamed to elected officials nationwide. In California, Governor Davis received $97,500 of Enron's $438,155 in contributions to state politicians. "Enron was smart because they went after people who they knew would make a difference," says Mary Kenkel, the former manager for federal affairs at the Edison Electrical Institute.

While awaiting incarceration, Enron officials have said that the company was proud of the role it played in energy deregulation, though in the aftermath of its bankruptcy filing and the implosion in California, it is now doing very little business in the states in which it once lobbied. "We were very active," says Mark Palmer, an Enron spokesman. "We helped open markets that needed to be opened. And there continue to be markets that need to be opened."

Indeed.

In fairness, Enron did take on some powerful interests, including local electrical monopolies, which are often the biggest campaign donors to statewide candidates. Enron representatives often came in cold, without any contacts in government or the lobbying community. To compensate, they lined up experts to testify, hired local lobbyists, and joined with consumer groups and some local utilities to present a united front for deregulation. And they wrote checks.

"They were basically strangers in the state capitols," says one industry observer. "They spent money to buy friends quickly and present their case quickly because they were up against utilities that had decades of local political relationships."

In California, for example, Enron spent more than $345,000 on lobbying, hiring former legislators and former utilities commission officials. As early as 1994, Jeffrey Skilling, the former Enron chief executive who then headed an Enron energy subsidiary, was testifying to utility commissioners that deregulation could save the state $8.9 billion—only half a billion dollars less than it eventually cost them.

Skilling told the regulators that California could "triple the number of police officers in Los Angeles, San Francisco, Oakland and San Diego. The stakes are huge and every minute that we delay bringing competitive markets to California allows the meter to keep ticking."

Well, the meter's still ticking; Mr. Skilling is still awaiting trial, and Californians are going to foot the bill.

The Texas Utilities Strike Back

We'll grant the obvious: Texas is nothing like California. For starters, there's plenty of electricity to go around. To quote Public Utilities Commission chairman Pat Wood, "We've got plenty of power in Texas. Even on the hottest day of the summer, during peak demand, we'll have more electricity than we'll need."

The Texas deregulation plan was the culmination of a legislative battle that began in 1995. Deregulation finally went into effect in 2002. While Enron did not get all it wanted, it did score a partial victory. Texans for Public Justice, a watchdog group, estimates that Enron's statehouse lobbying cost $535,000 to $945,000.

Enron hired 83 lobbyists in Texas, bought advertisements in local papers, and gave to local charities, including Laura Bush's Texas Book Festival. Incidentally, Governor Rick Perry of Texas, then lieutenant governor, received a $212,000 campaign contribution. At that point Perry served as president of the Texas Senate.

"Enron was unique because of the sophistication of their play," says Tom Smith, Texas director of a consumer group called Public Citizen. "It was all Enron, all the time. They helped craft the legislation. They gave to high-profile charities. They gave to both sides of the aisle. They'd hold fund-raisers for those they wanted to reelect. And they had the good ol' boy lobbyists go out after hours boozing and schmoozing."

About the only thing new and innovative in the competitive marketplaces nationwide was Enron's aggressive, take-it-or-leave-it approach. In New York, for example, where deregulation was enacted in 1996 but revisited by lawmakers in successive years, Enron made contributions to the state Republican Party in addition to George Pataki and hired a former state energy official, Howard Fromer, to lobby for it.

Enron chairman Ken Lay even called Governor Jeb Bush of Florida to push deregulation there. Florida ultimately did not pass energy deregulation, but lawmakers debated it for two legislative sessions.

Elsewhere, Enron's aggressiveness backfired. For example, the Enron style did not play well in Oregon, where Enron bought Portland General Electric and where partial deregulation was enacted in 1999.

"Enron came in like a house on fire, and we cooled them off," says Fred Heutte, energy coordinator for the Oregon Sierra Club. "Once they realized they wouldn't be allowed to do whatever they wanted, they lost interest in Oregon. Enron came in with a blatant attempt to roll the legislature and impress everyone with how important they were compared to podunk Oregon." Suffice it to say that residents of Oregon didn't appreciate the Texas attitude very much.

You have to look past the spectacular business failure to glimpse the true legacy of Enron's efforts. The fact is that consumer electrical markets have been made more competitive—even though Enron itself is no longer conducting business in many of those markets.

"For all of Enron's problems, they played an important role in opening up markets that were among the most fossilized in the country," says Robert Michaels, professor of economics at the University of California at Fullerton. "Whatever Enron did wrong, it spent a lot of money to achieve this."

And Enron wasn't the only Texas power company obsessed with exponential growth. Dallas-based Texas Utilities, or TXU, saw a real opportunity on the other side of the "big pond." In 1998, as Enron was making giant strides in various marketplaces, TXU made a historic $10.4 billion acquisition of Energy Group PLC, the largest utility in Britain at the time. TXU had been bidding against Portland-based PacifiCorp.

TXU's obvious strategy back in 1998 was to grow from a regulated utility to a deregulated power giant with operations on three continents.

This is a good example of what we call "the Michael Jordan Theory" as it applies to electric utility executives. Remember that Michael Jordan could not transition from being the best basketball player in the world to becoming one of the top 1,000 best baseball players, even though he tried for two years. Similarly, while these men and women may be good operators of monopolies, we are convinced they have much to learn about running a "real" business in a competitive marketplace.

TXU executives apparently sat in their plush Dallas conference rooms and convinced themselves that they could become "a powerhouse on three continents" in a competitive market environment. Just as Michael Jordan tried to convince himself that he could compete in a sport in which he had little experience, so TXU executives assured themselves that they knew how to play in a game with which they were almost totally unfamiliar.

How did things spin so badly out of control so quickly? In Britain, power prices dropped 40 percent after TXU's takeover of Energy Group PLC. The problem, as in California, was that the British company was contracted to buy much of the power it sells at above-market prices. The steady drain pushed TXU's once-shining European subsidiary to the brink of bankruptcy and forced it to renege on $200 million in debt to the supplying entity.

TXU lowered profit forecasts in October 2003, citing problems with the European operation. At that point, CEO Earl Nye said TXU would not need to cut ties with the UK operation; ten days later, the company put the British unit on the block. Nye said the move was necessary to appease credit-rating agencies.

Surely there is a "turnaround" award for such self-assured company pronouncements; if so, TXU ought to be in the running for top honors. At this writing, no court in London has declared the utility insolvent, but such a ruling may be only a matter of time. In the end, TXU will ante up more than $4 billion to get itself out of the European electric marketplace.

TXU played the game as well as it could for as long as it could, although it did so by deception and outright fraud. The company hid the truth from investors by "off-balance-sheet" tricks remarkably similar to those Enron had used. The bookkeeping shenanigans didn't help much; the utility has steadily lost ground. To make matters worse, TXU now faces lawsuits from two dozen disgruntled investors.

In essence, TXU executives could not translate their monopolistic business skills into a competitive business reality. When they could not succeed in a changing marketplace, they did just what old Samuel Insull did years ago: They resorted to a form of accounting alchemy. The result was rather predictable. As the old saying goes, truth will out.

One TXU analyst mourned what he called "a loss of credibility." He was more honest when he added, "Maybe a *lot* of credibility has been lost." In other words, the analysts have finally discovered that the folks at the helm didn't really know how to steer the ship. That should have been evident from the start; while education doesn't necessarily make the man, Nye's degree in engineering obviously failed to empower him for successful management.

According to the financial analysts, TXU must "execute what they've told us with regard to improving their balance sheet over the next year or so." What the analysts are saying is this: "You can't trust these guys, so we have to keep an eye on them."

TXU executives aren't exactly eager to blame themselves, but Nye did call himself and his colleagues "the victims of undue optimism." This may be the single worst excuse for incompetence ever recorded. Who forced, coerced, or enticed TXU executives to be overly optimistic? Simply stated, utility executives made poor decisions and tried to play power politics in a marketplace that they neither knew nor understood. Another midlevel executive was more candid: "Management sort of misled everybody in this situation." That this quote failed to raise eyebrows illustrates the manner in which business is conducted in the electric power industry. "Misleading" is commonplace in the industry.

Actually, TXU found a novel way to deal with the mess that Nye and his associates created. They went to Entergy and hired a new CEO. Hark back to the introduction, and you'll realize that the new TXU chief executive is the same individual we've identified as the finance officer who so challenged accounting standards with the Entergy alchemy (showing streetlights owned by the City of New Orleans as Entergy assets). We await the next chapter of this story with great anticipation!

The Light Bills Down in Georgia

Deregulation has yet to affect about half the country. Until it does—and even after, if the California and Texas experiences are any guide—it's business as usual for the utilities. What we find, time and time again, is that utilities continue to do pretty much whatever they want, regardless of whether the market they serve is competitive or not.

TXU executive Tom Baker says that because of his own company's experience and the Enron debacle, "the trend toward deregulation has slowed." Nowhere is that more evident than in Georgia, where deregulation has been effectively stalled by Georgia Electric's pledge to reduce rates by $300 million over three years. "The utility," one business observer dryly noted, "favors a slow approach to restructuring." In other words, restructuring will be completed around the time we are all dust.

Of course, restructuring isn't exactly a new process in Georgia. It happened once before, as the Civil War was winding down. In late December 1864, General Sherman offered the city of Savannah, Georgia—complete with captured cannon and cotton—to President Lincoln as a Christmas gift. Once the conquering Yankees had finished plundering the city, Savannah Electric took over. Just like the conquering armies from the North, utilities have almost always been able to do whatever they want.

The City of Savannah entered into a contract with our company in early 1996. Part of our audit would include street lighting. At that time, the city was spending about $200,000 per month on street lighting. Since 1994 Savannah Electric had offered the city no monthly itemization of power used or billed for street lighting. The city, like most of us, has always just written the check.

When we asked Savannah Electric to give me the documentation to support the monthly statements, the utility claimed not to know what we were talking about. "It's quite simple," we replied. "The city wants the bill to be right. Your bills should represent the power that was actually used."

The utility claimed not to understand our request.

However, it understood the request well enough to hire its own team of auditors, who came to Savannah and counted streetlights. "Here's the correct count," Savannah Electric finally told us, confident that its internal audit would end the matter. There was no apology for stonewalling us, and no offer to turn over additional documentation.

When we reviewed the utility's own audit, it became obvious to us that the utility was still billing for more streetlights than the total counted by the auditors the utility had hired.

But even when confronted with hard data, utilities have almost always been able to do whatever they wanted. We went out and counted streetlights ourselves and found that Savannah Electric had been billing the city for some 1,700 streetlights that did not exist. That 1,700-light figure was a monthly average; some months, the overcharges exceeded 2,000 additional lights, none of which were there.

All those "phantom" lights were costing the city about $20,000 every month. According to our calculations, Savannah Electric owed the city well over $1 million. We sat through several meetings before the utility finally admitted the obvious—it didn't have the streetlight count correct, and the billings had been in error. The utility's first offer to settle, however, was a paltry $100,000. That figure amounted to only about five months' worth of overcharges.

Needless to say, the city rejected the first settlement offer out of hand. Assistant city manager Bob Bartolotta asked the utility to calculate how much the city had been overcharged for the past twenty years. Of course, Savannah Electric didn't *want* to provide the twenty-year total. Instead, utility officials declared that their company was willing to resolve any billing disputes up to March 1994, when Savannah Electric's bookkeeping procedures had been changed. Anything prior to that date, according to the utility, was "not relevant to further negotiations."

We asked for clarification, failing to understand how something could be "not relevant" simply because bookkeeping methods had changed.

The utility reasoned that since the city was receiving itemizations prior to March 1994, the city was responsible for verifying those bills. The fact that the city had trusted the utility to furnish a correct bill meant the utility was off the hook—or so the utility believed. Prior to 1994, the City of Savannah had failed to catch Savannah Electric in the act of delivering what appeared to us to be fraudulent bills. By the time we requested the information that would prove the point, the utility's arbitrary statute of limitations had conveniently run out.

This sort of take-it-or-leave-it attitude has manifested itself in countless ways with countless utilities over the past several decades. Indeed, the inclination toward customer disservice is nothing new; it started with old man Insull's operation in Chicago more than a century ago. What is new, however, is the modern response.

What Georgia and other non-deregulated states are learning, albeit at a comfortable arm's length from the deregulation fray, is the same lesson learned the hard way in Ohio, California, and Texas: Even given adverse situations and circumstances, the utilities have the political clout and financial wherewithal to find a way to continue doing business as they always have. Enabled by regulators and elected officials, the power companies still manage to wind up on top. Deregulation can work, but only when those who actively aid and abet the electric utilities decide to put long-suffering consumers first instead.

Monopolistic Practices

Candidly, the topic of utility deregulation concerns others more than it does us. We are always asked about deregulation when we attend civic or social events. In the final analysis, deregulation, or the lack of it, does not have much of a bearing on our business or the work that is done.

Let us explain. In our business, we typically find five key areas in which electric companies have made errors, compounded mistakes, or committed outright fraud. Usually, when recovery for clients is made, it results from one or more of the following:

- Clients have been billed for electricity they did not use.
- Client utility metering has been manipulated in some way.
- The utility has sold the same electricity twice.
- Clients have been charged an incorrect rate for electricity used.
- Meter readings have been manufactured out of whole cloth.

Regulation and deregulation do not really have an impact in these five areas. After all, state or federal regulators will hardly be interested in some faked meter readings or in a business that has paid for power it did not actually use. The job of regulators is to look at the big picture. Our job is to look at the small things the regulators cannot be bothered to see.

In 1985 our business signed a contract with a major university here in Texas. We received copies of its bills and reviewed them. What we found was hardly surprising: several billing mistakes, some going back more than a year.

Until December 1983, the power company that supplied electricity to the university had never charged the school for peak demand, the portion of a calendar month during which the customer uses electricity at the highest rate. Peak demand is used to measure the greatest amount of electricity used over a specific period of time, usually fifteen or thirty minutes. Until that month, the university had always been charged only for actual kilowatt-hours used.

In December the utility added a peak demand tariff to its rate plan. As the new charge was implemented, many billing errors occurred.

When the utility discovered that our company had been hired by the university, communication slowed to a snail's pace. Just to show the company we meant well, we gave it a sop of sorts—a $25,000 undercharge that had resulted from a metering mistake. While the utility accepted the fact that a mistake had been made, the fact that the mistake was actually in its favor did not make much difference. Utility representatives quickly became obstinate and declined to return calls or answer letters.

Over the years, our company has developed what we call "fingerprints" of various electricity-using operations. A fingerprint represents a model or theoretical usage plan and tells us what could be considered normal or customary for a given type of firm. As a result of fingerprinting, we already know what to expect in terms of electricity usage, and we can quickly focus on any deviations.

Thanks to fingerprinting, we knew what usage patterns for a college campus should look like. After all, we had had more than a dozen colleges as clients. When we examined the usage profile for the client university, it became immediately obvious that something was amiss.

Electricity was delivered to the university through two service points. The school's distribution network was then used to conduct electricity to the various buildings around the campus. Each of these delivery points had two meters, and more than half of the university's electricity flowed through the four boxes. Strangely, these four meters displayed radically different consumption patterns—patterns that did not fit the typical college campus. Indeed, the patterns were quite a bit different from those of the smaller meters serving more remote parts of the university.

We wrote a letter to the electric company informing it that we thought the meters were functioning incorrectly. Three weeks later, the utility replied that the meters had been tested and found to be operating correctly.

We flew back to the university for a meeting and sat directly across the table from the associate director of the utility. We explained in great detail why the meters appeared to be wrong; the basic problem involved peak demand. In essence, the amount of electricity metered seemed far too high compared to the actual kilowatt-hours the university was using.

"The patterns we see here," we told the associate director, "might be created by a metal-fabricating operation. They are not usually created by a college or university."

Additionally, we mentioned that the consumption patterns revealed something strange. It looked as if all the equipment in every building on the campus was being turned on for a four-hour period on a daily basis, and then turned off—and that this was happening every single day.

The associate director knew better than to argue the point. Instead, he listened politely, then maintained that he and his colleagues wanted to take our findings under consideration. They would "get in touch" with us, we were told. We went back home and waited.

The "consideration" took about three weeks. Once again, our phone rang with a predictable message: The meters had been tested again and were found to be working correctly. We got back on a plane and flew out to the university to try to explain once again why there had to be an error.

This time we got a different response. "Who are you," the utility officials asked, "that you can sit in Waco, Texas, and tell us how to do our job?" This particular utility had been generating and distributing electricity for more than a century. Who were we to tell them what was right and wrong?

At this point, the utility took the show on the road, literally. At formal meetings on campus and at civic clubs, utility officials went out of their way to convince university administrators that we did not know what we were talking about. More than that, the officials said, we were causing problems and hurt feelings between friends in the community.

Of course, the university administrators began to have their own doubts. After all, everything the utility officers had been saying was true: We were just guys in Waco trying to tell a utility hundreds of miles away that its billing was incorrect. Suffice it to say that university officials became somewhat leery of our company.

We wrote another letter, had a third meeting, and got the same answer: "The meters are correct. There will be no adjustment made." Finally, we called the utility and arranged to bring an independent third party to the university for the purpose of testing the meters. "Bring him on," the utility told us.

The inspector met us at the university; he was unloading equipment from his truck when a utility representative showed up. He had something akin to a sheepish grin on his face. "Since we knew you were going to test these meters again, we tested them one more time." This would have been the fourth time the utility had examined the same two meters.

Funnily enough, the result of the test was strikingly different. The thirty-year-old meter was measuring peak demand at a rate two to three times higher than the actual demand. "Imagine that," we replied, knowing that, under certain circumstances, "Imagine that" is a remark of great import. Our inspector fell to testing the first meter and found it accurate. It should have been accurate, since the utility had repaired it the day before.

But we had other problems. Specifically, we were trying to figure out how a sticky meter could have affected the other metering points. Something was still wrong, so we began testing the other three meters. They all tested correctly, as they should have. All three had been replaced days before.

Later, utility representatives confessed that demand meters had never been installed at the three points. Back in 1983, when the power company was changing the way it metered electricity, they discovered that putting demand meters on all metering points was a costly endeavor. Instead of installing new meters, workmen found an old demand meter in their shop and installed it at one metering point. Meter readers took their monthly readings from this one defective meter and used the reading to estimate demand at the other three meters.

Those four meters generated quite a bill: The university was paying around $200,000 a month for that electricity. Although it made us quite uncomfortable, the utility apparently had no problem with estimating the large monthly sum. More astounding, the power company billed the university in good faith month after month—and the university paid the bills.

The month after the utility replaced the four meters, billing dropped by about $50,000. Since we had not had much luck with the utility, we asked the university vice president to go with us to see the utility president. The upshot of the meeting was this: The university wanted its money back. "These folks in Waco tell us you may owe us as much as a million dollars," the university vice president began, "and we really need that money."

The president of the utility apologized for the mistake but explained that this sort of minor error could have happened to anyone anywhere. "Thirty-four years ago, when I first went to work in the electric industry, my first job was dealing with guys like these who came to town and told everyone they were going to save them money. I'd have to follow those guys around and clean up the messes they made.

"But let me assure you," he said to the university vice president, "we don't owe you a million dollars. Those consultants of yours have just inflated the number because they get paid on contingency. We owe you some money and we want to pay it back, but we figure it's about $100,000 to $150,000.

"We've been doing this a long time and we know what we're talking about," the utility president continued. "These boys are just trying to get you all excited because they're thinking about their big fee."

Well, there was not a lot we could say. We did try to explain the logic inherent in the fact that when a power bill drops by $50,000 per month, and the situation has been ongoing for at least nineteen months, the amount of money owed the university has to be close to $1 million.

In the end, a third-party arbitrator settled the case for $1,044,000. All the while, the utility claimed the entire affair was a simple mistake.

We disagreed, of course. There was nothing simple about the utility's decision to estimate billing on a set of meters that generated $2 million a year for the utility. That decision was not an innocent mistake; it was quite possibly a criminal act. Still, many utilities allow the same thing to happen with client after client. When they get caught, power company officials are almost always of the opinion that they are being unfairly forced to pay for a simple mistake.

To Err Is Human, Right?

Another simple mistake involved a local client who owned a manufacturing operation. On his side of the electric meter, there were three feed lines—lines that supplied electricity to three separate manufacturing operations not owned by the same client. Our client was paying for every kilowatt of electricity that went through his meter—no matter where it ended up. Part of the electricity was used by his operation, but much of the power was siphoned off by the other operations that he did not own. The utility was not content merely to bill our client for electricity others were using; the power company metered the other three operations' usage as well, which effectively billed them twice for their power usage.

We had met this client at a Rotary Club gathering. He liked the idea of a utility audit and engaged us as a consultant. Because we work on a contingency basis, we get a part of whatever we recover for our clients, but if we recover nothing, we get nothing.

As we inspected our new client's plant, we noticed the subsidiary lines running from his meters to other buildings on the same industrial site. Obviously, the neighbors' electric lines were being metered through my client's meter first. The utility was billing all the businesses involved for the same electricity for which our client had already paid.

The etymology of the situation was simple enough to understand. Our client had purchased a plant that had originally been a part of a much larger facility. Instead of separating the power lines and billing his plant independently from the other buildings being used by other manufacturers, the utility billing department was told to just subtract the other manufacturers' usage from our client's bill. This billing adjustment hat trick worked for one month, then it never worked again.

All the while, our client was assured that the lines had been properly split.

We wrote a letter to the power company in an effort to straighten out the problem and obtain a refund for our client. Two days later our client got a phone call from a utility official. "Let me see if I can save you some money," the utility representative began. "How long have you had a contract with these guys?"

"About three months," our client replied.

"Well," the power company official responded, "we can show you in our file that we've known about this problem for more than six months. It's been our intention to tell you about this, correct the problem, and refund your money. You can just cut your consultants loose. You don't owe them anything."

Unbelievably, the power company representative thought our client would be thrilled to hear this. Understandably, our client became even angrier.

If our client dispensed with my services, he was told, the utility was prepared to issue a refund check for $78,000. Fortunately, our client is an ethical person. It was obvious to him that the utility was *not* operating under similar ethical constraints.

The utility called our client on August 5, but it claimed to have discovered the problem months earlier. Our client, in fact, had just mailed a $29,000 check in payment for his most recent electric bill. "Why didn't you call me earlier?" my client asked. The utility representative made some vague reply about being busy and things taking time.

The real truth is that the utility would probably never have bothered to tell our client about the problem if it had not first received the letter we wrote. Our client told the utility that we would continue to act as his agent. The utility stopped taking our telephone calls. The promised refund check never came.

Our client picked up a camera and photographed every piece of utility metering equipment on his site. He was afraid, as he put it, that the electric company "would come in and jerk the lines." He told every employee to immediately report any utility representative who set foot on plant property.

The very next week, one such representative showed up. An electric company workman called at the company gate and said he was there to separate the lines. Our client told him that the utility could do no work on the premises unless we were present to observe. The workman got back into his truck and left.

While we felt the utility's offer of a $78,000 refund was essentially accurate, our client also wanted to recover $4,000 in interest, our fees, and his attorney's fees as well. The total bill to the utility was $153,000.

Of course the power company never admitted to owing anything but the $78,000 overcharges. It even accused our client of charging usurious interest! But the utility also did not want to go to court; utility officials did not want to have this case exposed to public scrutiny. After all, they did not want every business owner out checking meters and wiring! Eventually, the utility settled the dispute by sending our client a check for $125,000.

It took the utility a year to spot the discrepancy, and until we wrote our letter, it had yet to split the lines or notify my client. "Getting a refund takes time," a utility official said, "and we don't want to contact a customer until we're ready to deliver a check."

If the customer should die of old age in the meantime, we thought to ourselves, the power company would be much better off.

In a truly deregulated environment, utilities would see the consumer as a sought-after commodity—as something of value to be courted, coddled, and appreciated. The pitfalls and shortcomings of deregulation are most easily proven by the attitude that utilities continue to manifest toward their customer base. While the utilities continue to find new ways to enrich themselves, ratepayers continue to get what they have always gotten—the shaft.

4

WHAT GOOD MAY COME

Liberation via Free Enterprise

The fundamental truth is that power companies will change the way they act only when they are *forced* to do so. The real problem lies in the fact that almost everyone in any position of authority—and that includes many state and federal officials—is beholden to the utilities in one way or another.

A wise man once said that anytime you allow politicians to do something for you, it's tantamount to giving yourself a blood transfusion from your right arm to your left arm—and spilling about half the blood on the way over. We wholeheartedly agree with that assessment; we've seen enough inefficient bureaucrats to last us a lifetime. Unfortunately, we've come to realize that only state and federal officials can muster enough power and support to force the utilities to change the way they do business.

In other words, it's up to the politicians to force the utilities to behave as they should. That's not to say that we must rely on our elected officials for the end result, nor are we advocating a socialistic takeover of the utilities themselves. Legislation at the state and federal level is, quite simply, the only force strong enough to bring about fundamental change in the way utilities operate. We believe that an appropriate end result will actually produce itself, once politicians properly set the stage. What government must then do is to put the machinery into motion.

Generally speaking, the problem with deregulation lies in the fact that the established utilities have found ways to manipulate the process. What is required is a government-leveled playing field. We've seen the apparatus of government move swiftly and confidently to create a fair and competitive environment in other arenas; there's no reason to think that the same system can't have the same effect on power companies as well.

The Southwest Airlines Story

The Wright brothers first flew at Kitty Hawk back in 1903. Thereafter, the most significant change in the history of powered flight came 73 years later, when the Civil Aeronautics Board asked Congress to dismantle the economic regulatory system and allow airlines in the United States to operate under market forces.

Congress obliged by passing the Airline Deregulation Act in 1978, which eased the entry of new companies into the business and gave them the freedom to set their own fares and fly whatever domestic routes they chose. Deregulation of the industry was quickly followed by new airlines competing for passenger business. The upstart competitors had the temerity to offer lower fares, and they were responsible for opening new routes and services to scores of heretofore unserviced cities.

Airline deregulation had a rough row to hoe. The airline industry faced the same situation as utilities face in deregulation. The growth in air traffic brought on by deregulation's first two years ended in 1981, when the country's professional air traffic controllers went on strike. After the strike was rather forcibly settled by the federal government, traffic surged again. Some 20 million new passengers flew each year in the post-strike period, and the airlines carried a record 466 million passengers in 1990.

But beginning in 1989, seemingly unrelated events started to undermine the economic foundations of the airline industry. The Gulf crisis and economic recession caused the airlines to lose billions of dollars. The airlines experienced the first drop in passenger numbers in a decade, and by the end of the three-year period 1989–1992 they had lost about $10 billion—more than had been made since the inception of the industry. Great airline names like Pan American and Eastern disappeared, while others, such as TWA and Continental Airlines, sought shelter from bankruptcy by going into Chapter 11.

Today, following the tragedies of September 11, 2001, the domestic airline industry is a low-cost, low-fare environment. Most of the major airlines have undergone cost restructuring, while others have sought the protection of Chapter 11 bankruptcy to restructure and reduce costs. Perhaps the only airline untroubled by recent catastrophes is Southwest, headquartered in Dallas, Texas.

When Southwest began flying in the 1960s, it flew only between cities in Texas—and was thus free from federal regulation. Later, when deregulation became a reality, Southwest was ideally positioned to occupy a unique niche in the marketplace. Southwest CEO Herb Kelleher has always emphasized two things: low costs and high spirits. Those who have flown Southwest Airlines—as we have—understand that the company's true success factor lies in its short-haul, high-frequency, low-fare strategy.

There are those who say that airlines simply compete for the same passengers and the winner is the one that succeeds in stealing the most from other airlines. That's not true with Southwest. "They take people out of Greyhound buses and their own automobiles," one industry analyst says. "Southwest occupies the nexus between two market spaces, the flying passenger who doesn't wish to pay for things that are not important, such as meals or reserved seating, and people who might otherwise consider driving to their destination."

In catering to this market niche, Southwest Airlines has forged a unique corporate culture. The nation's electric utilities might successfully emulate the airline by shifting their focus to providing worthwhile, cost-effective, reliable service to the millions of residential and other small power users nationwide. Unfortunately, the power companies haven't been motivated to change—at least not in that direction. By retaining the decades-old myopic focus on larger customers, the utilities have ignored the imperative to adapt to changing circumstances.

Enron and Other Failures to Adapt

And the results are telling. As businesses shrink and outsourcing sends more and more jobs overseas, the pool of large customers is beginning to shrink as well. All the while, there's an untapped resource of customers from whom just as much—if not more—revenue can be derived. Unfortunately, an investment of time and effort will be required to win this large resource over. We believe the kind of investment required is the single largest reason why the power companies have chosen to ignore the residential and small-customer base. Utilities are simply unaccustomed to the hard work involved in customer-focused service.

Failure to adapt seems endemic to the energy industry. Enron, in particular, failed because the company chose not to attempt to adapt to current market discipline. As a December 12, 2001, article in the *Wall Street Journal* noted:

The beginning of public education ought to be explaining what Enron actually was. The popular wisdom is that it was an energy trading company, and it certainly did buy and sell spot and futures contracts for gas and electricity, making markets in fancy financial derivatives designed to control risk.

But the more we inspect the record the more it looks as if Enron was in fact a huge hedge fund masquerading as a trading firm. Without telling its shareholders, Enron seems to have evolved into a high-risk investment firm operating with a huge pool of debt. We have nothing against hedge funds, as long as people know what they're dealing with.

The hooker in all this is that Enron was not just a hedge fund, but a publicly held hedge fund. The last hedge fund to fail in spectacular fashion was Long Term Capital, which was a private firm, which meant it didn't have the same obligations to disclose its financing to the public.

Those overseeing Enron obviously didn't possess the discernment of the *Journal*; and, of course, hindsight is always better than foresight. Economic growth and the demand for energy required Enron to adapt to a new way of doing business . . . and it didn't. Similarly, Enron failed to adapt to the industry-wide difficulty of obtaining the investment capital needed to expand the supply of energy. Indeed, all Enron did was find a new way to create money. Of course, the new methodology had a short shelf life and clearly lacked any semblance of ethical integrity. Fortunately for the energy market, dozens of companies were waiting to pick up the pieces when Enron fell apart. We believe that many large electric utilities, like Entergy, are following the Enron path to self-destruction.

The Harbinger of Change

Twenty years ago, when smoking was still allowed on airline flights, you'd see the smokers rush back to their cigarettes when the plane's wheels hit the ground. Tobacco has been replaced in modern times by the ubiquitous cellular phone. When the wheels touch down, the cell phones come out. Indeed, it's not uncommon for some particularly efficient cell phone users to make three or four calls between the time their long-haul flight lands and the time it arrives at the gate.

Such behavior would have been considered laughable and absurd twenty years ago, and we could not have foreseen the communications revolution from the same vantage point. Today, cell phones, pagers, and fax machines rule our lives, but they create such huge benefits, particularly in business and industry, that they are now considered an indispensable part of modern living. Something to think about: The phones we use today are much smaller (and, presumably, more fully featured) than the palm-sized communicators used in the original *Star Trek* series less than four decades ago.

This innovation in the telecommunications industry owes its existence, we believe, to the deregulation of the industry in the early 1980s, when the telecommunication monopolies were effectively broken up. All the advents of modern communication upon which we rely today—wireless Internet, digital cellular technology, and so on—were originally fostered by new telecom companies entering the just-deregulated marketplace. Before deregulation, entrepreneurs were squeezed out; the communications giants had no time for new ideas.

Former AT&T executives have publicly boasted that integrated-circuit technology was available to them as early as the 1940s and that communications advances such as the Internet could have been developed much earlier as well. That those who ran the telecommunications giant would admit—and even boast about—the availability of life-altering technology decades before it was actually marketed offers keen insight into the monopolistic mind-set of utilities.

In a word, AT&T had revolutionary technology at its disposal and did nothing with it. Content to make money the old-fashioned way, the communications monopoly carried on with business as usual for nearly four decades. Deregulation, not AT&T, was what brought innovation.

That these executives had such technology sitting on the shelf for decades and did nothing does not seem out of place to *them*. For us, however, it's the proof of the antiquated monopoly attitude. These people just don't get it. The notion of public responsibility or serving as a public trustee is clearly lost on them. "We made our money," their logic says, "and the dumb consumers didn't know what they were missing. No harm done!"

Similarly, the electric utilities have had no incentive to develop or evolve better ways of doing business. In a monopolistic environment, there was no motivation for entrepreneurs to invent and innovate in the area of electric power. Why? Quite simply, those with inventions to sell found that they could not market the technologies they had developed, because the power companies held a monopoly—and thus a stranglehold—on new technology reaching the marketplace.

Today, of course, burgeoning deregulation has brought about new and exciting areas of development. For example, great advances are now being made in the production of viable fuel cells. Who knows? Perhaps one day we will need no electric lines. The unsightly poles and wires that have become a common fixture in our lives will suddenly disappear; instead, an air-conditioner-sized device sitting beside our houses will produce all the electricity we need. Hydrogen would likely be the fuel; the only by-product would be water.

Believe it or not, such innovative power delivery is already a very real possibility. Before the technology can reach the marketplace, however, cost and reliability must be improved. Similarly, solar energy holds great promise. Eventually, an economical photovoltaic cell may revolutionize how we receive electric energy. Of course, none of us knows today what innovations lie ahead. We are certain, however, that innovation is the harbinger of change. Dramatic changes in the utility marketplace will be driven by innovation. And, for precisely that reason, innovation has been stifled for too long.

Adaptation and Survival

We believe that complete deregulation of electric marketplaces will trigger an avalanche of "gee whiz" technology. Old-line utilities must adapt, or like the dinosaurs of ages ago, they will die off, mourned by none and unloved except by a favored few. To avoid that fate, electric utilities must begin seeing customers, rather than profit, as the lifeblood of their business.

The future of electric power is not being decided in boardrooms. Instead, it is being created in garages, workshops, and laboratories by people with the innovative visions of Steven Jobs and Michael Dell.

Both innovation and ethical behavior are inexorably tied to this fundamental paradigm shift. It will be a difficult transition for the industry; utilities simply don't see customers as entities to be respected and appreciated. Our experience in West Hartford, Connecticut, bears out this point.

West Hartford is an interesting city, historic yet strangely provincial. West Hartford's power is supplied by Connecticut Light and Power. Like many other utilities, CL&P seems to operate under the illusion that it occupies a privileged position in West Hartford.

The City of West Hartford retained us in early 1996 to analyze recent utility billings. In early March we sent the first letter to CL&P requesting information on West Hartford's traffic signal billing. One account was being billed at the rate of about 30,000 kilowatt-hours per month for all the traffic signals in town. We were also curious about a couple of rate issues.

A month passed and we heard nothing from CL&P. Finally, a second letter elicited a reply, but it was less than helpful. Account executive Bill Majewicz responded that his utility "has no role in determining the loads of these unmetered traffic signals, but sets up billing, relying on the information provided by the customer or their agent. Many of the West Hartford units were established in the 1970's or earlier and it is doubtful that any of the information supplied is readily available today."

In a word, the billing for West Hartford's street signals was old and inaccurate, but the utility couldn't be bothered to check the usage and render an accurate bill. This is the issue of adaptation again: In any other business with a product or service to sell, the burden of demonstrating the need for or usage of that commodity falls upon the organization doing the selling. In this case, the utility clearly believed the City of West Hartford should be responsible for determining how much electricity the traffic signals actually used.

Of course, in a competitive environment, any business that let a customer go for two decades without reevaluating product usage would have long since lost the customer. CL&P, being an electric utility, faced no such danger.

The stalemate dragged on for nearly a year. Finally, during the last week of February 1997, we got a call from utility representative Byron Peart. "This whole thing is completely new to me," he admitted, and asked for documentation about the billing discrepancies we were investigating.

The documents we had, of course, *were from CL&P's own files.* Mr. Peart told us he wanted our copies of those letters. We asked if he had received a file with copies of the correspondence we'd exchanged with his office. He responded in the negative.

That file should have been fairly easy to find and transmit to Mr. Peart; it was more than an inch thick. Any other marketing organization would have had the information at hand—assuming it was at all interested in serving its customers.

But Connecticut Light and Power was obviously untroubled by such concerns. Seven weeks after our first conversation with Mr. Peart, we wrote a letter to city finance director Donna Sims. Our investigation had indicated the city was paying for many streetlights that did not exist and that CL&P had yet to provide documentation for traffic signal billing. CL&P, of course, maintained that the city bore the responsibility for CL&P's inaccurate billing.

Mr. Peart claimed that he still didn't understand the basic problem. By May 20, he had managed to send us documentation for one street-lighting account—without making mention of our request for documentation on nearly 30 more such accounts. We told Mr. Peart that his response was inadequate: "We are completely frustrated that we have explained so many times what is needed to verify the accuracy of billing . . . and we still cannot get it."

Our recommendation to the city was that it allow us to file a complaint with the Connecticut Department of Public Utility Control. When we eventually got the documentation we sought, we found that the City of West Hartford was paying for six dozen streetlights that did not exist. This represented an annual overcharge of about $15,000. CL&P owed the city more than $100,000.

In September 1997 we sent the streetlight billing review to Mr. Peart at CL&P. In January 1998, after only a four-month delay, Mr. Peart wrote back to tell us that our information was "under review."

That review must still be ongoing—we've never heard from Mr. Peart again.

The West Hartford story is emblematic of the necessity for a fundamental change in utilities' attitude. In order to adapt to changing markets and conditions, utilities must first recognize that they have a product to sell, and they must begin to view those who buy that product with something better than rank disdain and condescension. Thus far, deregulation has failed to produce such a paradigm shift.

The airline industry—and Southwest Airlines in particular—seems to have grasped this essential marketing truth and has managed to survive despite the consequences of treating customers like . . . well, customers. Indeed, some would argue that the industry has survived precisely *because* it now views passengers as a valuable commodity. We believe that a similar attitudinal change is critical to ending electric utility abuses and that the only way to force such a change is through elected officials at the state and national levels.

Of Bills and Tariffs

Before the tragedy of September 11, 2001, the World Trade Center complex was a beehive of activity centered around seven different buildings. The twin towers were only two of those structures; Six World Trade Center was located in the same complex that would become Ground Zero.

The New York Port Authority was supposed to provide electricity to 6 WTC, billing by what is called a declining block rate. For the first block of kilowatt-hours of electricity used, the client paid a given rate. For the second block, the rate declined, and so on. The work we had done with Six World Trade Center indicated there were on-site problems.

So we went to New York to visit the site and found our way to the building's basement. On one wall were three large meters. Farther to the right was a totalizing meter, from which the structure's bill should have been calculated. The totalizing meter was not being used. Through the use of the other three demand meters, the Port Authority was billing Six World Trade Center for artificially high maximum demand. We figured the overcharge to be in the neighborhood of 10 percent.

Additionally, Six World Trade Center had to work through the highest part of the rate structure three times per month because electricity coming into the structure was fed into the three individual meters. With all three meters added together, as they should have been, the building's total usage was enough to qualify for an even better rate.

Why was Six World Trade Center being billed as three separate buildings rather than as a single structure? Had the Port Authority simply chosen not to use the totalizing meter? We believed so. In fact, the Port Authority had claimed it did not own the meters, but the Government Services Administration, which owned the building, said that the meters did not belong to the GSA. Logically, the Port Authority bore responsibility for them.

We asked the Port Authority why the totalizing meter was not being used. "We don't do that," we were told. When we pressed the issue and pointed out that the totalizing meter had indeed been installed, the response changed but remained substantially vague: "That's just the way someone wanted it," the utility replied.

As we dug more deeply into the rates being charged by the Port Authority, we discovered something equally odd. Six World Trade Center and the surrounding buildings purchased their electricity from the Port Authority, which was its own political subdivision. The Port Authority was supposed to charge for electricity at the prevailing local rate, meaning that it could not exceed what the local power company would charge for the same service. The Port Authority computed its bills exactly the same way Consolidated Edison did, with appropriate tariffs, and so on. In fact, the Port Authority used Consolidated Edison's tariff for billing purposes.

Included in each bill was a Gross Receipts Tax of 8 percent, which was identical to the same tax imposed by Consolidated Edison. The difference lay in the validity of the tax itself; the Port Authority, as a political entity, did not pay the tax—in effect, it was tax exempt. The same could not be said for Consolidated Edison. ConEd, being a corporate entity, was subject to the tax. Rather than forwarding the tax to the City of New York, the Port Authority just put the 8 percent surcharge into its rather large pocket.

As it turned out, the General Services Administration chose not to go after the Port Authority for overcharges. The GSA's regional counsel would not let us challenge the Port Authority on the Gross Receipts Tax issue.

Eventually the Port Authority began using the totalizing meter to determine billing for Six World Trade Center. By that time, however, the GSA had overpaid the Port Authority more than $15,000 per month for several years. While the electric bill did decrease by about 10 percent, the Gross Receipts Tax continued unabated, and the GSA dutifully paid the bill.

The World Trade Center complex was like a city unto itself. Over the years, we have found actual municipalities being billed incorrectly; the difference usually lies in the fact that cities want their money back.

One municipal client contacted us about a billing problem at the city hall. Officials had contacted the provider, Gulf States Utilities, three times about electric bills that seemed too large. Gulf States representatives had visited the city hall twice and proclaimed the bills to be entirely accurate. The problem, they told our client, was that the city hall building itself was poorly designed and thus used an inordinate amount of electricity.

After about fifteen minutes on-site, we were able to document that the Gulf States bills were incorrect. All it took was a simple look at the meter, which was pegged on the highest setting. The city had been billed for usage of 450 kilowatts of demand for eighteen of the last twenty-four months. The proof of the billing error lay in the fact that the transformer at the city hall could handle only 300 kilowatts. 450 kilowatts could not have gone through a 300-kilowatt transformer without damaging the transformer itself.

The Gulf States representatives had bothered to look at neither the meter nor the transformer; the utility had simply assumed the equipment was functioning properly. And, of course, it was not. To the utility's credit, it settled the claim immediately when we pointed out the problem.

Not all municipalities are quite so observant. One afternoon in the summer of 1998, we were driving north on Stemmons Freeway in Dallas. The gray concrete towers of the downtown area loomed ahead of us, the eight-lane freeway fairly teemed with vehicles heading in and out of the Metroplex. We pulled off the highway and parked near a utility pole at the bottom of the exit ramp. Parking well outside the lane of traffic, we got out and walked to the utility pole to examine the meter mounted on it.

The meter box was a rusting vestige of its former self. While the meter itself appeared to be intact, it was obviously nonfunctional and had been for quite some time. Shards of glass from the meter housing lay on the ground, and creeping vines grew through the meter mechanism. But we could still read the meter number: 127 052 507.

We flipped through the pages of the latest electric bill dutifully paid by the Texas Department of Transportation. Meter 127 052 507 appeared on page seventeen, along with the latest reading and the current charge. God Himself couldn't have read that meter—but Texas Utilities had.

At least, its bill stated that it had.

The Inherent Challenge of Re-regulation

As Enron and other energy marketers found out the hard way, the price of producing electricity can fluctuate from minute to minute. In the deregulated marketplace, the price paid for power can be even more volatile. This makes deregulation seem a risky proposition, but in reality, deregulation and competition are both absolute imperatives and long overdue.

We've already seen that the move toward deregulation has been sparked (pardon the pun) by changes in fuel technology. Lower fuel prices have made it feasible to generate electricity with only a minimal capital investment in equipment. Additionally, the integration of individual utilities into regional power systems has effectively increased the size of the market that power generators can now service.

But just because electricity generation is easier now does not mean that the process is risk-free, or that there's no room for government regulation. While we believe the days of co-opted regulators have largely run their course, government still has much to do to ensure that electricity is reliably and fairly marketed.

For power companies to regain our trust, we believe the process of deregulation or re-regulation must follow three rather simple steps.

First, competition must become the standard rather than the exception in every utility market.

Local governments should accelerate the speed with which their markets are deregulated—although one can hardly blame them, given the situation in California and elsewhere, for not being in a particular hurry to get the job done. If deregulation forces providers out of business, elected officials have a crisis on their hands. After all, in times of electricity shortages, not just a few homes and offices are affected. As the massive power outages have shown us in recent years, everyone suffers equally when the lights go out.

Granted, government regulation of the electric utilities can still lead to outcomes that are even less attractive to consumers than the abuses already occurring. We believe, however, that market power is usually self-correcting, and that a short-run exercise of market power usually attracts new competitors to the market. Even the threat of other competitors can effectively discourage old-line utilities from pushing their prices higher.

What this means for the utility industry as a whole is very simple: Power companies must bow to market pressure like every other industry. The free enterprise system does work! Consequently, given time, deregulation will work—but only with certain other controls in place.

However, the free enterprise system is based on the Golden Rule: Do unto others as you would have them do unto you. All customer service philosophies we've seen are based on a version of "give and you will be given to." Both are essential attitudes that must be forced upon the electric utilities.

Free enterprise is property-based. Because it is impossible to give what you do not own, businesses must own their means of production. We have no problem with that, so long as the means of *distribution* remain in the public domain. The model for electric distribution lines should be akin to our highway system, in that anyone can use the network to transact business between locations of their choice. In that respect, utilities ought to learn to act as public trustees. Clearly, thanks to the history of the industry, this is a role with which the power companies have had altogether too little experience.

Zealous regulation has its place. Our airwaves, for example, are guarded by the Federal Communications Commission. Regulation of broadcast and television in the United States is a model for other countries to follow, largely because broadcasters willingly adhere to a code of community service.

What we have seen in our business, time and time again, is examples of electric utilities unhindered by the demands of community service or competitive forces. Freedom from the constraints of service and the lack of competition breed inefficiency, disinterest, inaccuracy, and blatant arrogance. What Samuel Insull did for Thomas Edison's electric company a hundred years ago has not only hindered innovation throughout the utility industry and bilked millions of investors but also has rendered a monumental disservice to the industry's customers.

As we have seen, the only beneficiaries of the monopolistic practices that have resulted have been the utilities themselves. However, the benefits the utilities have gained have been dubious at best. Had old man Insull not "insulated" Edison's company from competition, the emerging electric utility industry might have become so much more than it is today.

Second, electric utilities should pledge to serve in the public interest as public trustees before being given access to distribution apparatus. And the good intentions inherent in that pledge must be carried out.

Without such a pledge, access to means by which electricity is distributed would be denied. Regulators would assume only the role of enforcing the public service aspect; the free market and competition should dictate how much power companies can charge for their product.

This system would work today, were it not for the fact that electric utilities have been so accustomed to behaving so badly for so long. Additionally, the utilities have powerful allies in willing politicians at every level who have been effectively co-opted by the influence of utility cash. As we have seen, these co-opted officials will often look the other way while the utility does whatever it pleases.

The real success of deregulation has been largely overlooked: Essentially, deregulation freed up distribution apparatus for use by any company with the means and desire to compete in a given marketplace. What we propose is a governmental monitoring of the use of that apparatus for the public good. To continue with our highway analogy, licenses and permits would be denied to those companies with the electric equivalents of an unsafe driving record and worn-out vehicles.

Why not just go back to the old days of monopolies and regulation? Regulation, in our view, will always create more problems than it solves. Additionally, regulators—who are public officials themselves—are, like their elected counterparts, always subject to being co-opted. We also believe that the free enterprise system is dependent upon freedom to sustain itself. Since successful entrepreneurs typically defy the conventional wisdom, the life of the enterprise itself requires individual and organizational freedom. For better or worse, the utilities never experienced that freedom under governmental regulation.

Third, electric utilities and those who operate them must become entrepreneurs in their own right.

Because entrepreneurs must serve and collaborate with others in order to succeed, electric utilities must be operated by people of character. Both are essential to the act of putting one's fate into the hands of others unknown in a free market characterized by voluntary choice. The simple test of whether those currently in control of the power companies have the required character lies in whether they accept or reject this new role.

We believe we have adequately demonstrated that many electric utilities and many of those who operate them lack the strength of character necessary to be effective entrepreneurs. Clearly, we've met our share of regulators who lacked the strength of character to deny the utilities what was not theirs. But among the many selfish and greedy individuals, there are bound to be a few ethical entrepreneurs who live to serve and serve to live.

Like Diogenes with his lamp, we go out into the world every day looking for the honest utility executive. Implausible though it may seem, we have indeed met our fair share of honest utility executives, although our business naturally brings to our attention those who are anything but honest. But we remain perpetually hopeful that even among the corrupt, a few good men and women may still be found. All we have to do is patiently search for them.

Once a new deregulated, competitive system is up and running, we believe the free enterprise system will take over. The system will then be self-perpetuating of its own accord.

Is Ethical Behavior Too Much to Ask?

If America's electric utilities would just begin doing what they're supposed to do—behaving, for lack of a better word, like the decent corporate citizens they claim to be—we would be out of business overnight. Indeed, were utilities honest and forthright, there would be no need for what we do.

You may call us foolishly optimistic, but we believe that the utilities can be taught the necessity of playing by the same rules under which the rest of us operate. For so long, the power companies have gotten away with whatever they wanted. Changing that behavior now requires more than just catching them in the act from time to time.

Utility apologists argue that power companies *are* trying to behave according to a standard of some sort. Unfortunately, it seems to us that the accepted standard of ethical behavior is never met. Over the past few decades, the borders of acceptable behavior and decorum have been bent, stretched, and tortured by the utilities, and the end result is that the power companies honestly believe their operations are not subject to the same ethical constraints as every other business up and down the block.

We believe that ethical behavior should be expected from every enterprise—even from individuals and organizations operating on the edge of delusion. And again, our ignorance of the classic utility attitudes and mind-set may be partly to blame for the failure of the power companies to play by the rules.

Every American ought to care about utility mismanagement and malfeasance. Our collective lack of concern is the death knell for our efforts to change the behavior of the power companies. Because they know how little we concern ourselves with what they do and how they do it, the electric utilities know they have free rein to ride roughshod over us. And they've done so with wild abandon.

That's a sad thing to say about the companies that supply the very lifeblood of our society. Even sadder is the realization that we have freely given them the upper hand. The problem, as with so many social, political, and economic issues, is that few of us can agree on what's really wrong and how best to go about fixing the situation. And until we reach some kind of consensus, the utilities will win by default.

Having been on the front lines in the war against utility cheating, we are convinced that the real problem lies within the deeply ingrained habits of thought that the utilities themselves possess—and in the passive acceptance of those habits by consumers. These are the essential attitudes that we seek to change, and we are convinced that they *can* be changed. Unless these habits of thought and action are altered, utility cheating will go on unchecked. Utilities should be held just as accountable as any other type of business. The fact that they provide power and other services that have become essential to American society should not dissuade any of us from taking a long, hard look at the bills we pay each month. Similarly, auditors and regulators need to look hard at utilities' financial statements—especially in the areas of property records and asset valuations!

Doing anything less is tantamount to giving up the battle before the first shot is fired. To allow the utilities to remain virtually unaccountable means that we have chosen to abdicate control over a substantial portion of our livelihood. The monthly allocation to utilities is an extraordinarily broad and prevalent spending pattern, affecting everyone from the individual consumer to the state and national governments.

The real problem is that almost all of us have abdicated that demand for accountability for too long. Utilities have become conditioned to getting away not only with vast sums of our money but with our misplaced trust as well. Power companies and other utilities have literally squandered our faith in them—they have squandered the trust we've placed in them by (wrongly) assuming that they desired to tell the truth and to do what's right. Until all of us assume responsibility for managing our utilities and service providers and holding them accountable, we all remain at risk.

We can repair utility billing practices, and we can hold utilities just as accountable as they hold us. After all, power companies and other service providers have no special authority over us; indeed, we should have some sort of authority over *them*. Because the potential for unintentional fraud and intentional misapplication is so large, utilities should be held to a higher standard than, say, the local convenience store.

For too many years, it's been the other way around. In no uncertain terms, utilities have demanded accountability from *us*. Have you tried *not* paying your electric bill lately? In return, the power companies have offered precious little accountability of their own. If we want to change the system, we must first change the concept that the utilities are somehow deserving of special treatment or privilege.

To justifiably expect some degree of accountability, we must first understand the inner workings of the system we are scrutinizing. Most of us have at least some rudimentary grasp of how government works; of course, the basic principle of tax-and-spend isn't hard to comprehend. Most of us have been through the educational system in one format or another; thus, we can bring our own personal knowledge to bear when we demand accountability from our schools. And we can demand accountability from businesses because we usually know something about the product we're purchasing.

As things stand, few industries are held *less* accountable than our electric utilities. Indeed, most of us expect more accountability from our office supply store than we do from our electric company. That's because electricity, the system that delivers it, and the utilities themselves are deep mysteries to most Americans.

Frankly, we despair of finding another way to help Americans learn more about their utility providers and the way in which they work. That is, after all, the reason for this book, but we harbor no illusions about the task in which we have involved ourselves for so many years. The fact is that most of us just want to flip a switch and see the immediate result. We have more important things to do than study how the switch works, how the electricity got to the switch, and how the power company bills us for the trip.

For the utilities, our ignorance is their bliss. Unburdened by a watchful consumer base and freed long ago from the shackles imposed by zealous regulators, the electric utilities can pretty much do whatever they want. The past behavior of the power companies we've battled is by far the best evidence of this de facto license—in a very real sense, the utilities have a license to plunder.

Deregulated or not, the current system has been hard-wired for greed, outright fraud, and eventual failure. We believe it's time someone pulled the plug.

978-0-595-35744-4
0-595-35744-X